P9-DEZ-864

NUMERICAL METHODS
FOR UNCONSTRAINED OPTIMIZATION

NUMERICAL METHODS FOR UNCONSTRAINED OPTIMIZATION

Edited by

W. MURRAY

National Physical Laboratory
Teddington, Middlesex

1972

ACADEMIC PRESS · LONDON AND NEW YORK

ACADEMIC PRESS INC. (LONDON) LTD.
24/28 Oval Road,
London NW1

United States Edition published by
ACADEMIC PRESS INC.
111 Fifth Avenue
New York, New York 10003

Library of Congress Catalog Card Number: 72–84447
ISBN: 0-12-512250-0

PRINTED IN GREAT BRITAIN BY
THE WHITEFRIARS PRESS LTD., LONDON & TONBRIDGE

Contributors

C. G. BROYDEN, *Department of Computer Science, Essex University, Colchester, Essex.*

R. FLETCHER, *Theoretical Physics Division, Atomic Energy Research Establishment, Harwell, Berkshire.*

M. J. D. POWELL, *Theoretical Physics Division, Atomic Energy Research Establishment, Harwell, Berkshire.*

W. H. SWANN, *Imperial Chemical Industries Corporate Laboratory, P.O. Box 13, The Heath, Runcorn, Cheshire.*

W. MURRAY, *Division of Numerical Analysis and Computing, National Physical Laboratory, Teddington, Middlesex.*

Preface

This book is based on a joint IMA/NPL conference which took place at the National Physical Laboratory on the 7th and 8th January 1971. The decision to publish the proceedings of the conference was not taken until after the conference and this has delayed publication. The authors have taken advantage of this delay to include more recent research material.

In keeping with the IMA policy to hold expository conferences, a decision was taken to follow the Keele research conference with an expository conference on some topic of optimization. In choosing the topic to be unconstrained optimization two points were considered. Firstly, two days had to be an adequate time to cover all relevant points, and secondly, the material presented should not become obsolete during the next few years. In the past decade very significant strides had been taken in unconstrained optimization and rather than becoming less relevant much of this past work was becoming of increasing interest.

This volume has two objectives: the first is to give a comprehensive and detailed survey of the current numerical methods available for unconstrained optimization and the second is to provide the reader with a framework to help him follow future developments. Optimization is essentially a practical tool and one principally used by non-mathematicians; in contrast, most research papers in optimization are written in a style that is only intelligible to a mathematician. This volume attempts to bridge this gap. Little mathematical knowledge is assumed on the part of the reader. The first chapter outlines basic theory necessary to understand the remaining chapters and an appendix gives some results from linear algebra with which the reader needs to be familiar.

Most of the chapters are concerned, as the title suggests, with the description of *methods* for the solution of unconstrained optimization. To solve almost any practical problem it is necessary to utilize a computer. Knowing a method of solution could be a long step from having a computer program based on that method that will successfully solve problems. Some advice on the pitfalls of implementing the different methods are described in Chapter 7, and in Chapter 8 the sources of various ALGOL and FORTRAN programs are listed.

The state of the art is still a long way from providing a single all-purpose computer program that can effectively and efficiently solve even the majority of current problems. It is likely to remain this way for some time. The resourceful problem-solver is, therefore, forced to acquire some background knowledge of the subject in order to choose the most effective algorithm to solve his problem. The author of a computer program is often faced with the conflicting interests of different potential users. Consequently there is always scope for the intelligent and informed user to improve the performance of an algorithm on a particular problem.

May, 1972 W. MURRAY

A Glossary of Symbols

The common notation used by the authors of this book is set out below.

1. The problem of concern can be stated as follows

P1.
$$\underset{x}{\text{minimize}} \{F(x)\}$$
$$x \in E^n,$$

where $F(x)$ is a nonlinear function of the variables

$$x = \begin{bmatrix} x_1 \\ x_2 \\ \cdot \\ \cdot \\ \cdot \\ x_n \end{bmatrix}$$

and E^n denotes n dimensional Euclidean space.

2. g is the $n \times 1$ gradient vector of $F(x)$, that is

$$g_i = \frac{\partial F(x)}{\partial x_i}, \qquad i = 1, 2, \ldots, n.$$

3. G is the $n \times n$ Hessian matrix of $F(x)$, that is the (i, j)th element of G, $G_{i,j}$ is given by

$$G_{i,j} = \frac{\partial^2 F(x)}{\partial x_i \partial x_j}, \qquad \begin{array}{l} i = 1, 2, \ldots, n. \\ j = 1, 2, \ldots, n. \end{array}$$

4. $B^{(k)}$ is an $n \times n$ matrix that is a kth approximation to G.
5. $H^{(k)}$ is an $n \times n$ matrix that is a kth approximation to G^{-1}.
6. $x^{(k)}$ is the kth approximation to $\overset{*}{x}$, a minimum of $F(x)$.
7. $F^{(k)} = F(x^{(k)})$.
8. $g^{(k)}$ is the gradient vector of $F(x)$ at $x^{(k)}$.
9. $y^{(k)} = g^{(k+1)} - g^{(k)}$.
10. $s^{(k)} = x^{(k+1)} - x^{(k)} = \alpha^{(k)} p^{(k)}$,
where $\alpha^{(k)}$ is a scalar and $p^{(k)}$ is a direction of search.

11. A superfix ^{T} on a matrix or vector denotes transpose.
12. C^r denotes the class of functions whose rth derivative is continuous.
13. $\|y\|$ denotes an arbitrary norm of y.

Apart from a few hopefully obvious exceptions capital letters are used to denote matrices, lower case letters to denote column vectors and Greek letters to denote scalars.

Contents

Contributors v

Preface vii

Glossary of Symbols ix

1. Fundamentals 1
   W. MURRAY

2. Direct Search Methods 13
   W. H. SWANN

3. Problems Related to Unconstrained Optimization 29
   M. J. D. POWELL

4. Second Derivative Methods 57
   W. MURRAY

5 Conjugate Direction Methods 73
   R. FLETCHER

6. Quasi-Newton Methods 87
   C. G. BROYDEN

7. Failure, the Causes and Cures 107
   W. MURRAY

8. A Survey of Algorithms for Unconstrained Optimization 123
   R. FLETCHER

APPENDIX: Some Aspects of Linear Algebra Relevant to Optimization 131

References 135

Author Index 141

Subject Index 143

1. Fundamentals

W. MURRAY
National Physical Laboratory

1.1 Introduction

The principle purpose of this chapter is to give the general theoretical background necessary to understand the remaining chapters. In addition to this, topics of a general nature are discussed to prevent repetition by the other authors.

Although any function $F(x)$ must have a least value, the value is not necessarily finite. It could even be that $F(x)$ does not take its least value in E^n. A simple example of this is when $F(x)$ is a linear function other than a constant. What most people are interested in when presented with $P1$ is a solution $\overset{*}{x}$ of a certain character and this leads us to the following definitions.

Definition (1)

A point $\overset{*}{x}$ is said to be a strong local minimum of $F(x)$ if $F(x)$ is defined on a δ-neighbourhood of $\overset{*}{x}$ and there exists an ε, $0 < \varepsilon < \delta$ such that

$$F(\overset{*}{x}) < F(x),$$

for all points such that

$$0 < \|\overset{*}{x} - x\| < \varepsilon.$$

Definition (2)

Let $F(x)$ be defined on a δ-neighbourhood of $\overset{*}{x}$. The function $F(x)$ is said to have a weak local minimum at $\overset{*}{x}$ if $\overset{*}{x}$ is not a strong local minimum but there exists an ε, $0 < \varepsilon < \delta$ such that

$$F(\overset{*}{x}) \leqslant F(x),$$

for all points such that

$$0 < \|\overset{*}{x} - x\| < \varepsilon.$$

Definition (3)

A point $\overset{*}{x}$ is said to be a global minimum of a function $F(x)$ if for $x \in E^n$

$$F(\overset{*}{x}) \leqslant F(x).$$

For an arbitrary function there is no guarantee that such an $\overset{*}{x}$ exists since $F(x)$ may take its least value at a limit as $\|x\| \to \infty$.

There are a number of alternative expressions used to describe the different types of minima. The more common of these are listed below.

TYPE OF MINIMUM	ALTERNATIVE EXPRESSIONS
Strong Local Minimum	Strong Relative Minimum Proper Relative Minimum
Weak Local Minimum	Improper Relative Minimum
Global Minimum	Absolute Minimum

Figures 1.1 and 1.2 illustrate the various types of minima.

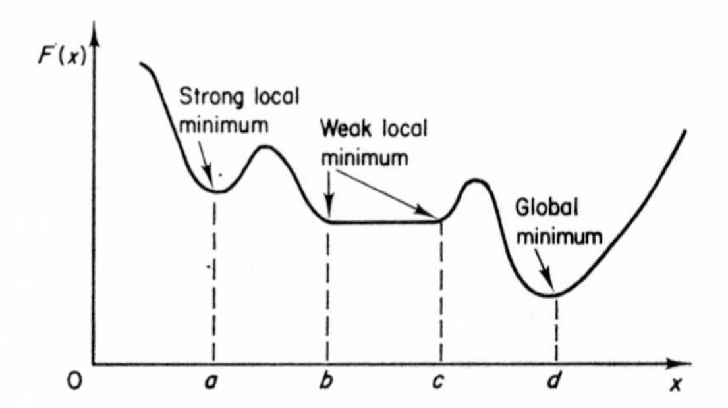

FIG. 1.1. A function of a single variable.

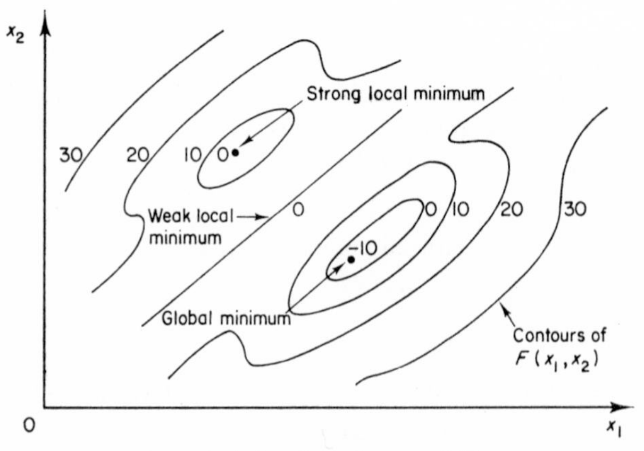

FIG. 1.2. A function of two variables.

When it is not necessary to distinguish between a strong and weak local minimum only the term local minimum will be used.

1.2 Necessary and sufficient conditions for a minimum

Nothing constructive can be said about the behaviour of $F(x)$ at $\overset{*}{x}$ unless $F(x)$ has certain continuity properties.

Consider for example a function that is discontinuous, in particular one such that

$$\lim_{\lambda \to 0} \{F(\overset{*}{x}) - F(\overset{*}{x} + \lambda h)\} \neq 0,$$

where h is an arbitrary unit vector,
and λ is a scalar.

An example of such a function is illustrated in Fig. 1.3. It is unlikely that an algorithm which is efficient on well-behaved functions would find the minimum at $x = 1$. Consequently algorithms are designed to find local minima of only a specific class of functions. Usually an algorithm will

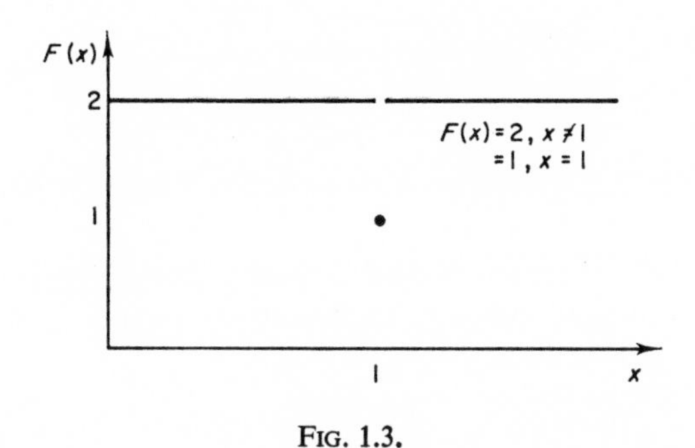

FIG. 1.3.

assume or utilize properties of the class of functions it has been designed to minimize. All the algorithms described in this book except for those given in Chapter 2 assume

$$F(x) \in C^1 \quad \text{for } x \in E^n. \tag{1.2.1}$$

Most of them also assume

$$F(x) \in C^2 \quad \text{for } x \in E^n. \tag{1.2.2}$$

Such functions are common. It is of interest to examine the conditions on $F(x)$ at $\overset{*}{x}$ for functions for which either (1.2.1) or (1.2.2) hold.

Definition (4)

A point $\hat{x}$ is said to be a stationary point of $F(x)$ if

$$\hat{g} = \nabla_x F(\hat{x}) = 0.$$

A necessary condition for $\overset{*}{x}$ to be a local minimum of $F(x)$ if $F(x) \in C^1$ is that $\overset{*}{x}$ is a stationary point.

Definition (5)

A point $\overset{*}{x}$ is said to be a saddle point of $F(x)$ if $\overset{*}{x}$ is a stationary point but not a local minimum or maximum.

Figure 1.4 illustrates a saddle point. The derivation of the term saddle point is from the two dimensional case.

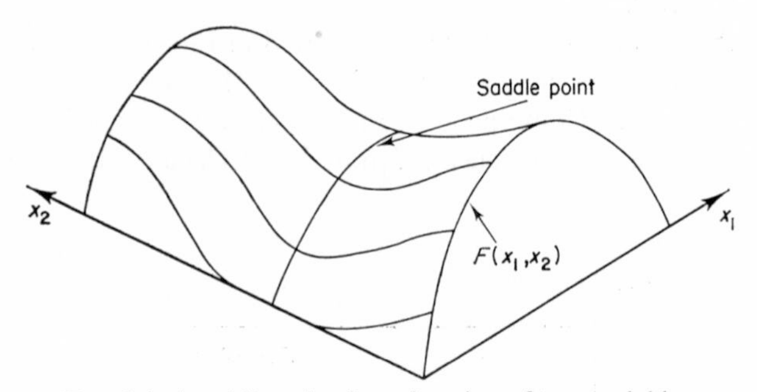

FIG. 1.4. A saddle point for a function of two variables.

Unless the class of functions to which $F(x)$ belongs is further restricted little can be said about sufficient conditions on $F(x)$ at $\overset{*}{x}$. For the class of functions

$$F(x) \in C^2,$$

$\overset{*}{x}$ is a strong local minimum of $F(x)$ if

$$\overset{*}{g} = \nabla_x F(\overset{*}{x}) = 0,$$

and G, the Hessian matrix of F(x), is positive definite at $\overset{*}{x}$. For the same class of functions a necessary condition for $\overset{*}{x}$ to be a local minimum is that G is positive semidefinite at $\overset{*}{x}$. These conditions on the derivatives of $F(x)$ are apparent if we consider the Taylor expansion of $F(x)$ about $\overset{*}{x}$. If $F(x) \in C^1$ then

$$F(\overset{*}{x}+\varepsilon y) = F(\overset{*}{x})+\varepsilon y^T\overset{*}{g}+O(\varepsilon^2),$$

where y is an $n \times 1$ arbitrary unit vector
and ε is a scalar.

For ε sufficiently small it is clear that

$$F(\overset{*}{x}) \leqslant F(\overset{*}{x}+\varepsilon y)$$

implies

$$y^T \overset{*}{g} = 0 \quad \text{for all } y.$$

This in turn implies

$$\overset{*}{g} = 0.$$

The Taylor expansion of $F(x)$ about $\overset{*}{x}$ for $F(x) \in C^2$ reduces to

$$F(\overset{*}{x}+\varepsilon y) = F(\overset{*}{x})+\tfrac{1}{2}\varepsilon^2 y^T G y + O(\varepsilon^3).$$

If G is indefinite then there exists some vector, say $\hat{y}$, for which

$$\hat{y}^T G \hat{y} < 0,$$

hence for ε sufficiently small

$$F(\overset{*}{x}) > F(\overset{*}{x}+\varepsilon \hat{y}).$$

It follows that G must be at least positive semidefinite for $\overset{*}{x}$ to be a local minimum and that for G positive definite, $\overset{*}{x}$ is a strong local minimum. For functions with higher continuous derivatives and G positive semi-definite it is possible to give additional conditions to distinguish between the different type of stationary points. Since it is impractical to verify these conditions computationally we do not pursue the matter further. A discussion of these higher-order conditions is given in Chapter 2 of the book by Gue and Thomas (1968).

1.3 Quadratic functions

It is unlikely that a minimization algorithm would be required to find the minimum of a quadratic function since it will be shown that this is equivalent to solving a set of linear simultaneous algebraic equations. Regardless of this, the behaviour of an algorithm on a quadratic function is important in optimization. One reason for this is that it is the simplest type of function that can be minimized since all linear functions apart from $F(x)$ = constant are unbounded below. A more important reason is that the behaviour of a minimization algorithm on a quadratic function is indicative of its behaviour in the neighbourhood of the solution when $F(x) \in C^2$.

Consider the quadratic function

$$F(x) = \tfrac{1}{2}x^T G x + b^T x,$$

where G is an $n \times n$ symmetric matrix
and b is an $n \times 1$ vector both of which have constant elements. Then

$$g = \nabla_x F(x) = Gx + b.$$

From the necessary conditions for a minimum given in §1.2 it follows that

$$G\overset{*}{x} + b = 0.$$

Although $F(x)$ may not possess a local minimum, it must possess a stationary point if b lies in the range of G. When G is non-singular there is a unique stationary point. If G is positive definite then

$$\overset{*}{x} = -G^{-1}b,$$

is a strong local minimum. If G is positive semidefinite then $F(x)$ has a weak local minimum if b lies in the range of G.

When $F(x)$ has a weak local minimum say $\overset{*}{x}$ then

$$y = \overset{*}{x} + z,$$

where z is any vector lying in the null space of G, is also a weak local minimum.

An important concept when minimizing a quadratic function is that of conjugacy. It can be shown that there exists an $n \times n$ matrix P of rank n such that

$$P^T G P = D,$$

where D is a diagonal matrix. There are many such matrices, for instance, the columns of P could be the eigenvectors of G. If we make the transformation of variables

$$x = Py,$$

then

$$F(y) = \tfrac{1}{2} y^T D y + b^T P y,$$

so that $F(y)$ is a separable function of the variables $(y_1, \ldots, y_n)$. The minimum of a separable function can be found by minimizing with respect to the variables individually. This is identical to minimizing $F(x)$ along the vectors $p_1, p_2, \ldots, p_n$, where p_j is the jth column of P. The directions $p_1, \ldots, p_n$ are called a conjugate set of directions.

1.4 Convex functions

It is not usually known or possible to determine whether a function is convex. Despite this, convex functions and their properties are important in optimization. The reason for this is that convergence of a minimization algorithm can often be proved for convex functions. Since convex functions are so rare this would not be particularly valuable were it not for the fact that many functions are convex in the neighbourhood of a local minimum.

Definition (6)

A function $F(x)$ is said to be strictly convex if for any two points, say x and y,

$$F(\lambda x + (1-\lambda)y) < \lambda F(x) + (1-\lambda)F(y),$$

for $0 < \lambda < 1$.

Definition (7)

A function $F(x)$ is said to be convex if for any two points, say x and y,

$$F(\lambda x+(1-\lambda)y) \leqslant \lambda F(x)+(1-\lambda)F(y),$$

for $0 < \lambda < 1$.

The difference between a convex function and a strictly convex function is illustrated in Figs 1.5 and 1.6.

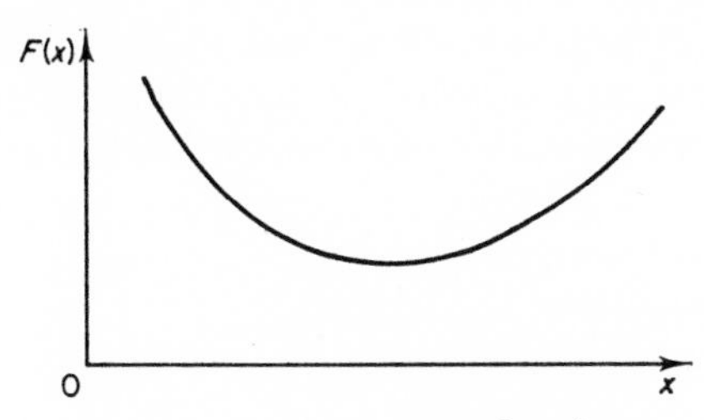

FIG. 1.5. A strictly convex function.

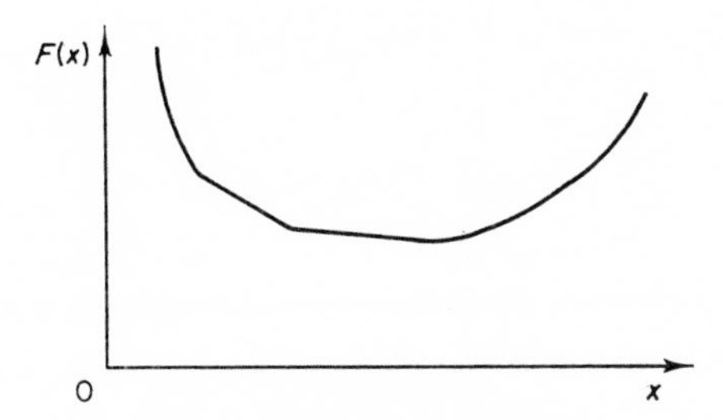

FIG. 1.6. A convex function.

If $F(x) \in C^2$ and is strictly convex, G is positive semidefinite at every point in E^n. Further, if $F(x)$ is a quadratic function and G is positive definite, then $F(x)$ is strictly convex. If $F(x)$ is a convex function and has a strong local minimum then this is unique and is the global minimum. Should $F(x)$ have a weak local minimum and $\overset{*}{x}$ and $\overset{*}{y}$ are two such minima then

$$F(\overset{*}{x}) = F(\overset{*}{y})$$

and all the points on the line joining $\overset{*}{x}$ and $\overset{*}{y}$ are weak local minima.

1.5 Functions of a single variable

The problem of minimizing a function of a single variable occurs repeatedly in some algorithms that minimize a function of n variables. Within each iteration nearly all these algorithms determine a parameter, say α, which either minimizes or reduces $F(x(\alpha))$. Typical of such algorithms is the following iterative process.

k*th iteration*

(i) A direction of search $p^{(k)}$ is determined.
(ii) A scalar $\alpha^{(k)}$ is determined which minimizes $F(x^{(k)}+\alpha p^{(k)})$ with respect to α.
(iii) $x^{(k+1)} = x^{(k)}+\alpha^{(k)}p^{(k)}$.

This process is illustrated in Fig. 1.7. The procedure of finding a minimum or a reduced value of $F(x)$ along $p^{(k)}$ is usually referred to as a linear search. Although there are a large number of algorithms designed to perform a linear search there are basically only two approaches to the problem.

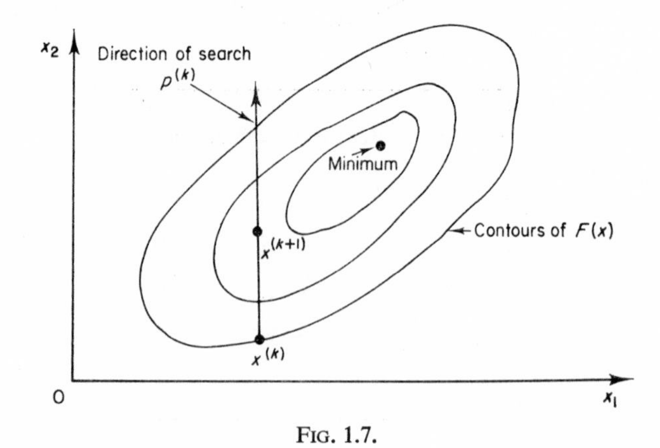

FIG. 1.7.

Method 1: Function comparison

The essential feature of this approach is that function values are used only in comparison tests. Typical procedures of this type are the methods of Fibonacci, Bisection and Golden Section. We shall assume that the function has a single minimum and that initially we have four values of α, say $\alpha_1 < \alpha_2 < \alpha_3 < \alpha_4$ such that the minimum lies in the interval (α_1, α_4). These initial values can be found by starting at one point and taking steps of increasing size until a function value larger than the previous value is obtained. If this occurs after the first step the minimum has been bracketed otherwise the sign of the step is reversed and the procedure repeated. By observing three consecutive points it is possible to determine whether the minimum lies in the interval (α_1, α_3) or (α_2, α_4). The function is evaluated at a new value of α lying in the same interval as the minimum. Again we have four points that bound the minimum but now the interval between the end points has been reduced. Successive iterations further reduce the interval until a satisfactory approximation to the minimum has been found.

The difference between the various methods is in the choice of α for which the new value of the function is made. It can be shown that for a *specified* number of function evaluations Fibonacci search is optimal (Kiefer, 1957) in the sense that it gives *a priori* the smallest interval in which the minimum lies. In most problems, however, it is not possible to specify in advance the number of function evaluations required.

A method based on Fibonacci search which is almost as efficient and does not require the number of function values to be specified is the method of Golden Section. Initially we choose four points (see Fig. 1.8) α_1, α_2, α_3, α_4 that bound the minimum and satisfy the equations

$$\alpha_3-\alpha_1 = \alpha_4-\alpha_2 = \beta(\alpha_4-\alpha_1)$$

where

$$\beta = 2/(1+\sqrt{5}) = 0{\cdot}618\,034\ldots$$

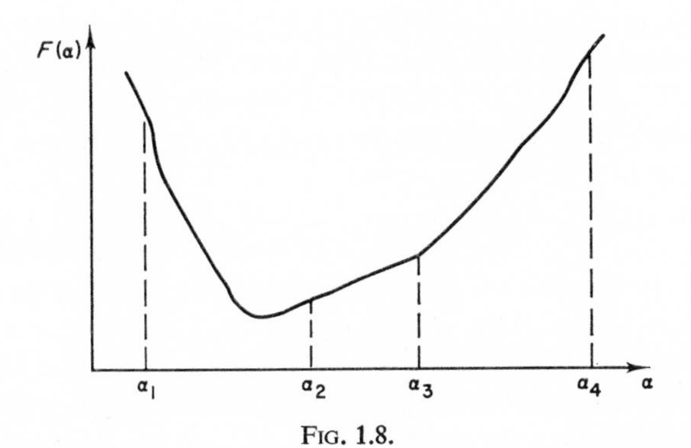

FIG. 1.8.

By testing the relative values of the functions $F(\alpha_1), \ldots, F(\alpha_4)$ it is possible to determine in which of the two intervals (α_1, α_3) or (α_2, α_4) the minimum lies. Since these intervals are of identical length no advantage is gained if we assume that the minimum lies in the interval (α_1, α_3). The function is then evaluated at a new point, say α_5, where

$$\alpha_5-\alpha_1 = \alpha_3-\alpha_2.$$

The special significance of β is that

$$\alpha_3-\alpha_5 = \beta(\alpha_3-\alpha_1).$$

If the points are now renamed

$$\bar{\alpha}_1 = \alpha_1, \qquad \bar{\alpha}_2 = \alpha_5,$$
$$\bar{\alpha}_3 = \alpha_2, \qquad \bar{\alpha}_4 = \alpha_3,$$

the position is similar to the initial stage except that

$$\bar{\alpha}_4 - \bar{\alpha}_1 = \beta(\alpha_4 - \alpha_1).$$

Hence each iteration, which involves one function evaluation, reduces the interval in which the minimum is known to lie by a factor β.

The disadvantage of this and similar methods is that they fail to utilize all the information available. The information they disregard is the quantity by which the functions differ at the various points.

An approach that utilizes this information is that of function approximation.

Method 2: Function approximation

The essential feature of this approach is that along the search direction p the function $F(x)$ is approximated by a simple function $\hat{F}(\alpha)$. The functions are related in that at a fixed number of points the function values and possibly the derivatives are identical. The function $F(x)$ is evaluated at the point at which the minimum of $\hat{F}(\alpha)$ occurs. This new function value is then used in forming a new approximating function. Obviously $\hat{F}(\alpha)$ must be a function whose minimum can easily be determined and so it is normally chosen to be a second or third order polynomial.

The first derivative of $F(x)$ along p is found by evaluating g and forming the inner product $p^T g$. Similarly second derivatives involve evaluating G and forming the product $p^T G p$. Consequently although first derivatives are sometimes worth evaluating it is unusual for a linear search procedure within a general minimization algorithm to use any higher derivatives.

We shall again assume that along the direction p, $F(x)$ has only one minimum. Suppose we are given three points along p and the function evaluated at these three points. The stationary point $\hat{\alpha}$ of the second order polynomial passing through these points is given by

$$\hat{\alpha} = \frac{1}{2}\frac{(\alpha_2^2-\alpha_3^2)F_1+(\alpha_3^2-\alpha_1^2)F_2+(\alpha_1^2-\alpha_2^2)F_3}{(\alpha_2-\alpha_3)F_1+(\alpha_3-\alpha_1)F_2+(\alpha_1-\alpha_2)F_3} \tag{1.5.1}$$

where α_1, α_2, α_3 are the steps along p at which the function is evaluated and F_1, F_2, F_3 are the respective function values.

Given two points α_1 and α_2 with function values F_1, F_2 and derivatives g_1 and g_2, a stationary point $\hat{\alpha}$ of the third order polynomial passing through these two points and having the specified derivative values is given by

$$\hat{\alpha} = (\alpha_2 - \alpha_1)(1-(g_2+\gamma-\eta)/(g_2-g_1+2\gamma)), \tag{1.5.2}$$

where

$$\gamma = (\eta^2 - g_1 g_2)^{\frac{1}{2}},$$

$$\eta = 3(F_1 - F_2)/(\alpha_2 - \alpha_1) + g_1 + g_2.$$

The above stationary point is the one which lies in the interval (α_1, α_2) if the minimum of $F(x)$ along p lies in this interval. Assuming $\alpha_1 < \alpha_2$ then the minimum lies in the interval (α_1, α_2) if $g_1 < 0$ and $g_2 > 0$.

The next question that arises is which of the current points the new point should replace. If the old values bracket the minimum the new value will lie within this interval, consequently the interval in which the minimum is known to lie will be reduced. The function value at the new point is not necessarily lower than all the values that were used to predict this point.

If the point corresponding to the largest function value is replaced, the minimum need no longer lie in an interval containing the new set of points. Consequently neither of the formulae (1.5.1) or (1.5.2) can be relied upon to give valid results. An alternative is to ensure that the set of points always brackets the minimum. Unfortunately the rate of convergence of the resulting process can become very slow.

A procedure based solely on approximation methods will either be unreliable or inefficient. The way round this dilemma is to ensure reliability and efficiency by mixing the strategy of the two basic methods for linear search. The next method to be described does precisely this.

In the "mixed" method the new estimate to the minimum $\overset{*}{\alpha}$ is given by some approximation formula unless this lies outside predetermined bounds. In the next iteration the points corresponding to the lowest function values are used. If the minimum is known to lie in a certain interval then the bounds are based on this interval. A typical iteration utilizing gradients is illustrated in Fig. 1.9. The minimum is known to lie between α_1 and u_1. The points α_1

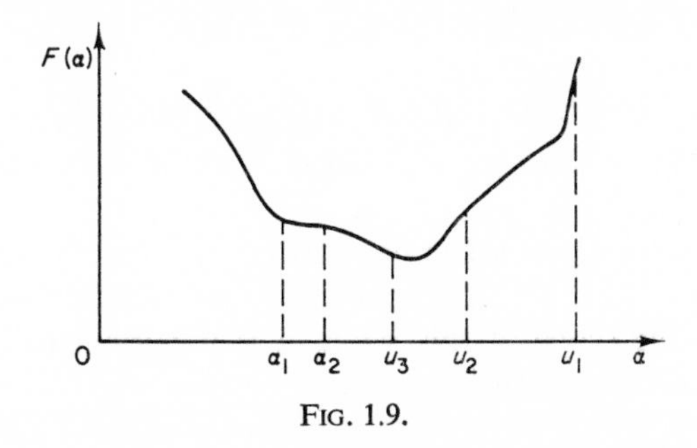

FIG. 1.9.

and α_2 are used to predict the minimum but this prediction will only be accepted if it lies in the interval (α_2, u_2). If it lies outside this interval then the new estimate is u_2. Suppose that u_2 is the next estimate, then the best two points will remain the same and the prediction in the next iteration will also lie outside the bounds. The minimum is now known to lie in the interval (α_2, u_2) so a new bound is constructed from this to give u_3. The location of the bounds u_2, u_3, etc., is usually taken to be at the point bisecting α_2 and the best known upper bound. There are more elaborate procedures which base the bound on the intervals (α_1, α_2) and (α_2, u_1) (Gill *et al.*, 1972).

If no interval including $\overset{*}{\alpha}$ is known then the bound is based on the interval (α_1, α_2). Suppose in Fig. 1.9 the function values and first derivatives were only known at α_1 and α_2 then an upper bound

$$u_1 = \alpha_2 + \beta(\alpha_2 - \alpha_1)$$

would be placed on the prediction of $\overset{*}{x}$. The scalar β is usually taken to be 2 or 4.

This section is not intended to be a detailed description of a linear search algorithm but merely a guide to the principles behind such algorithms. There are many numerical difficulties involved in performing an accurate linear search, some of which are discussed in Chapter 7.

1.6 The minimization of functionals

The function $F(x)$ described in $P1$ is a particular example of a functional, that is, a transformation from some linear space Y into the space of real scalars. In $P1$ the linear space Y is E^n with distance, inner-product, etc., suitably defined. The familiar concepts of differentiation, distance and inner-product can be defined on most linear spaces. Consequently nearly all the algorithms described in this book can be extended, at least theoretically, to the minimization of more general functionals, for example, to the following problem,

$$P2 \qquad \underset{x(t) \in D}{\text{minimize}} \left\{ V(x(t)) = \int_0^1 F(x(t), t)\, dt \right\},$$

where

$$D = \{x(t) : x(t) \in W' ; x(0) = a, x(1) = b\},$$

W' is the linear space of real valued absolutely continuous functions defined on $[0, 1]$,

and $F(x(t), t)$ is some prescribed nonlinear function of $x(t)$ and t.

Although algorithms for minimizing the type of functional appearing in $P2$ are beyond the scope of this book, it is important to realize that the relative merits of a particular implementation of an algorithm on a computer will change when the type of functional changes. A good example of this is provided by conjugate gradient methods (Chapter 5) which are much easier to adapt to solve $P2$ than are the quasi-Newton procedures (Chapter 6).

2. Direct Search Methods

W. H. SWANN
Imperial Chemical Industries
Corporate Laboratory,

2.1 Introduction

Almost all numerical methods for determining the optimum of a given non-linear objective function are iterative and starting from a given initial estimate for the solution they proceed by generating a sequence of new estimates, each of which represents an improvement over the previous ones. The different procedures are characterized by their strategies for producing this series of improving approximations, and direct search methods are those whose strategy is based on the comparison of values of the objective function only; thus such methods make no use of any of the derivatives of the function.

Although most direct search methods have been developed by heuristic approaches, some of them have proved extremely effective in practice, particularly in applications in which the objective function was non-differentiable, had discontinuous first derivatives, or was subject to random error.

This paper describes some of the more useful of these direct search methods, omitting those which generate conjugate directions as they are considered by Fletcher in Chapter 5 of this volume. The review begins with those methods which assume that the minimum is known to lie within a given region of E^n, such as the extension to n dimensions of the univariate search technique based on the Fibonacci numbers. A generally more successful approach is to define a set of n independent direction vectors and to direct the search according to the results of explorations along them. A number of such techniques are discussed, some of which, for example the procedure proposed by Hooke and Jeeves, retain the original directions throughout the search, while others attempt to generate an improved set after each iteration, as in the method due to Rosenbrock. Finally, methods in which the search is guided according to the result of comparing the magnitudes of the function values at the vertices of some geometric figure are considered. This approach is based on the use of statistical designs and originates from the Evolutionary Operation technique devised by Box to improve on-line the productivity of industrial processes.

2.2 Bounded methods

If the required minimum is known to be within a finite area defined by upper and lower bounds on each of the independent variables, i.e.

$$u \geqslant \overset{*}{x} \geqslant l,$$

then a number of very simple optimization procedures can be devised. For example, one of the most immediately obvious approaches would be to generate at random a sequence of points lying within the specified area with the one corresponding to the least value of the function taken to be the required minimum. Most high level computing languages provide a facility for obtaining random numbers according to a fixed distribution so that this type of search is very easily implemented on a computer. Unfortunately it is also very inefficient, for assuming a flat probability density function is used in generating the random points so that a particular trial point in the area of interest is as likely to occur as any other, then the number of function evaluations required to be 90% confident that the uncertainty of variable x_i has been reduced to an amount e_i can be shown (Spang, 1962) to be approximately

$$2{\cdot}3 \prod_{i=1}^{n} \left(\frac{u_i - l_i}{e_i}\right).$$

The efficiency of the method can be improved by carrying out only a proportion of these function evaluations, selecting a reduced area of search around the best points, making a random search in this new area, further reducing the search region, and so on, but the method still remains generally unsatisfactory.

A more systematic variation of this bounded approach is to set up a grid over the area of interest and to evaluate the function at each node of the grid with the minimum assumed to be that point corresponding to the smallest function value. This method, too, is very simple to program, but very inefficient. Thus to achieve the same accuracy as in the above random search requires

$$\prod_{i=1}^{n} \left(\frac{u_i - l_i}{e_i} + 1\right)$$

function evaluations which in general is considerably less than the number needed by the random method, but which also increases exponentially with the number of variables, and the method is of little practical value.

The principal disadvantage of the random search and grid methods is that the whole search strategy is predetermined and no scope for learning is allowed. All of the remaining methods to be described involve the comparison of each new trial point with one or more of the previous ones, and the generation of subsequent trials depends upon the result of that comparison. Thus the methods possess a memory and they attempt to compile information about the function so that the search can be directed towards the minimum, resulting generally in a considerably better performance.

2.3 Generalized Fibonacci search

A useful method for locating the minimum of a unimodal function of a single variable known to lie within fixed bounds is that based on the Fibonacci numbers and described by Box, Davies and Swann (1969), and it has been extended by Krolak and Cooper (1963) to apply to multivariate unimodal functions.

In the one-variable search the interval of uncertainty is successively reduced by comparing the function values at two interior points of the interval. Thus if it is known that $a \leqslant \overset{*}{x} \leqslant b$, then trial points $x = c$ and $x = d$ are selected where $a < c < d < b$ and a reduced interval chosen according to the relative values of $F(c)$ and $F(d)$; if $F(c) < F(d)$ then assuming a unimodal function the minimum must lie in the interval $[a, d]$ so that the interval $[d, b]$ can be discarded, but if $F(c) > F(d)$ then the reduced interval is $[c, b]$. The procedure is then repeated using this reduced interval and continues until the interval has been sufficiently reduced to ensure that the minimum has been located to the required accuracy. The efficiency of such a strategy depends upon the positioning of the trial points $x = c$ and $x = d$, and the Fibonacci search chooses them such that if $[a, d]$ is the reduced interval then $x = c$ is also a trial point of it, and similarly $x = d$ is a trial point of $[c, b]$ when it becomes the reduced interval. Thus for each successive reduced interval one trial point is already available and only a single function evaluation is necessary for each iteration after the first.

This procedure can be extended to deal with multivariable problems in the following manner. Consider first the two dimensional case where the problem is to minimize $F(x_1, x_2)$ subject to $a_1 \leqslant x_1 \leqslant b_1$ and $a_2 \leqslant x_2 \leqslant b_2$. Two points on the x_1 axis are chosen, $x_1 = c_1$ and $x_1 = d_1$, and then two one-variable Fibonacci searches are made in the x_2 direction, i.e.

$$\min_{a_2 \leqslant x_2 \leqslant b_2} \{F(c_1, x_2)\} \quad \text{and} \quad \min_{a_2 \leqslant x_2 \leqslant b_2} \{F(d_1, x_2)\}.$$

This yields two function values to associate with $x_1 = c_1$ and $x_1 = d$, respectively so that they may be compared, the interval of uncertainty with respect to x_1 reduced and the procedure repeated. For example, in the case illustrated in Fig. 2.1, the function value associated with $x_1 = c_1$ is greater than that associated with $x_1 = d_1$ so that $[c_1, b_1]$ becomes the new reduced interval for x_1.

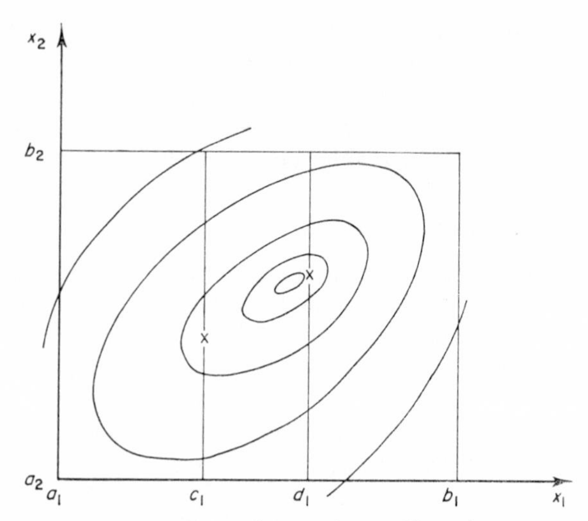

FIG. 2.1 Fibonacci search in two dimensions.

The extension to higher dimensions follows in the same way so that for example in three dimensions each "function evaluation" in the x_1 direction involves a two-dimensional search as described above. Thus if α_i function evaluations are required to reduce the interval of uncertainty with respect to x_i, then the total number of evaluations required is $\prod_{i=1}^{n} \alpha_i$ which in general is very considerably less than the number required in the random and grid methods. However, for problems involving more than two or three variables it is still not a very practical method, although it has been reported to work well on small dimensioned problems in which the function exhibited sharp, narrow valleys.

2.4 Alternating variable method

An approach which has produced a number of more useful direct search methods is to define a set of directions which can then be used to explore the parameter space. The most simple strategy of this form is that known as the *alternating variable method* which consists of minimizing with respect to each independent variable in turn. Thus starting from the given initial approximation variable x_1 is altered, with $x_2, \ldots, x_n$ held constant, until a minimum of the objective function is located whereupon x_1 is fixed and x_2 is explored in the same way, and so on until x_n has been searched. Hence the search proceeds parallel to each co-ordinate direction in turn, changing direction each time a minimum is located, and for functions whose contours are hyperspherical or elliptical with the axes parallel to the co-ordinate directions, the method is very efficient since if the linear searches are exact then the exact minimum will be obtained. Figure 2.2 illustrates the performance of the method on a function of two variables for which the minimum lies at the centre of a set of concentric circles.

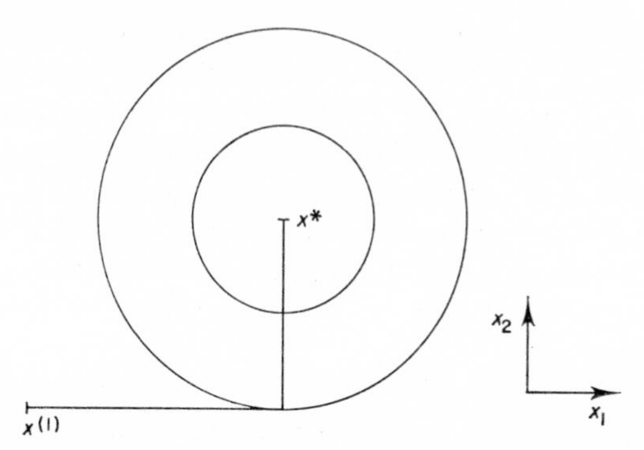

FIG. 2.2. Alternating variable method for a function of two variables having circular contours.

However, in general there will be interaction between the variables so that the function contours are skew with the co-ordinate axes. In this event the procedure must be extended by searching along x_1 again after x_n has been explored and cycling continuously around the directions in this manner until no progress can be made along any direction when it is assumed that the minimum has been located. As shown in Fig. 2.3 this generally results in a search which progresses along the principal axis of the contours by

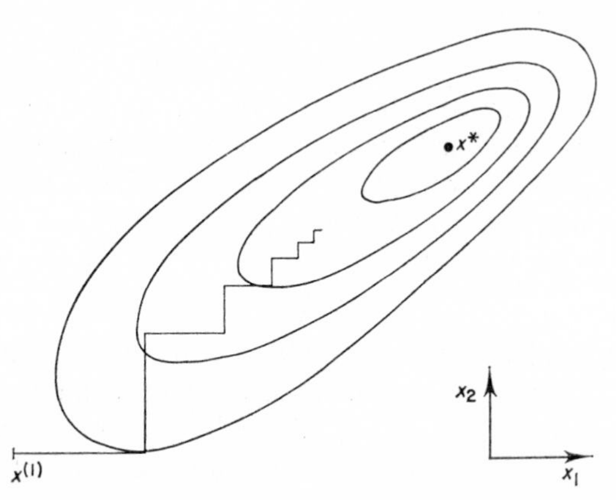

FIG. 2.3. Alternating variable method for a function of two variables having skew elliptical contours.

means of a series of ever-decreasing steps, so that in most practical cases the method is very inefficient and it has been noted that the inefficiency tends to become worse as the number of variables increases.

2.5 Pattern search

Clearly to be efficient a search technique must define and pursue a direction which lies along the local principal axis of the contours of the function. The *pattern search method* due to Hooke and Jeeves (1961) attempts to do this using a combination of exploratory moves and pattern moves to generate a sequence of improving approximations $x^{(1)}, x^{(2)}, x^{(3)}, \ldots$ called base points. Exploratory moves examine the local behaviour of the function and seek to locate the direction of any sloping valleys present; pattern moves utilize the information yielded by the explorations by progressing along any such valleys.

In an exploratory move each co-ordinate direction is examined in turn in the following way. A single step is taken along the direction (by adding an increment to the relevant co-ordinate) and is considered successful and the new value of the co-ordinate retained if the function value has not been increased. If the step fails it is retracted and replaced by a negative step which is either retained or retracted according to whether it succeeds or fails. When all n co-ordinate directions have been investigated the exploratory move is complete. The point arrived at as a result of this procedure, which may or may not be distinct from the point from which the move originated, is generally termed a base point.

A pattern move consists of a single step from the present base point, that step having both magnitude and direction of the line joining the previous base point to the current one. Thus it is a step from the present base point $x^{(k)}$ to the point $\bar{x}$ where

$$\bar{x} = x^{(k)} + (x^{(k)} - x^{(k-1)}).$$

The search begins by considering the initial estimate for the solution as the first base point $x^{(1)}$ and making an exploratory move from it to find a new base point $x^{(2)}$ where $F(x^{(1)}) \geqslant F(x^{(2)})$. Then, on the assumption that the exploration has determined a useful downhill direction, a pattern move is made to the point $\bar{x}$ where

$$\bar{x} = x^{(2)} + (x^{(2)} - x^{(1)}).$$

It is of course unlikely that the initial exploration located exactly the best direction to follow, so the pattern step is accepted, at least temporarily, regardless of whether it succeeds or fails and an attempt is made to improve the pattern direction by exploring about the point $\bar{x}$ to obtain the point $\hat{x}$, where of course $F(\hat{x}) \leqslant F(\bar{x})$. The function value at $\hat{x}$ is then compared with that at the present base $x^{(2)}$, and if $F(\hat{x}) \leqslant F(x^{(2)})$ then a new downhill direction $(\hat{x}-x^{(2)})$ has been located so $\hat{x}$ becomes the next base point $x^{(3)}$, and from it is made a new pattern move followed by an exploratory move. The procedure continues in this way with each pattern step followed by an exploration to ensure that the pattern direction continues to be a downhill one. However, eventually the situation where $F(\hat{x}) > F(x^{(k)})$ for some base point $x^{(k)}$ will occur indicating that the usefulness of the current pattern direction has been exhausted, so the pattern plus exploratory moves that were made from $x^{(k)}$ are discarded and a new exploration made about $x^{(k)}$ to determine a new pattern direction.

The above procedure reaches an impasse when, a pattern direction having been abandoned, all steps about the base point in the ensuing exploratory move fail so that no new direction has been defined. This is taken to indicate that the base point is either close to the minimum or in a sloping valley whose sides are too steep to allow a direction along the valley to be located using the current exploration step sizes. The remedy in either case is to reduce the step sizes and carry out a new exploration. Convergence is assumed and the search terminated when the exploration steps have been reduced below some specified limit.

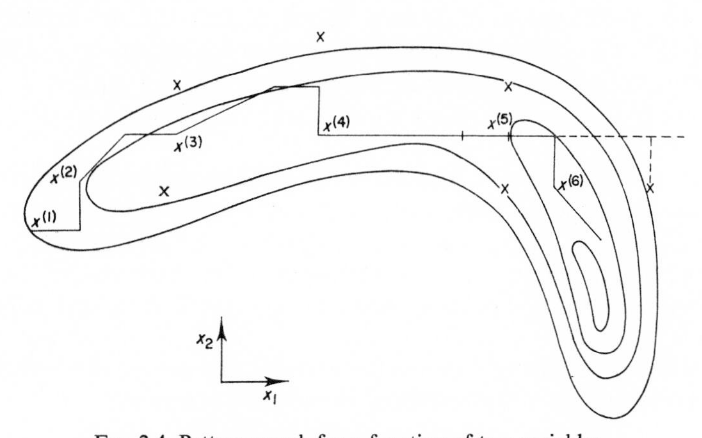

FIG. 2.4. Pattern search for a function of two variables.

The performance of the algorithm on a function of two variables is depicted in Fig. 2.4 in which the base points are numbered in the order in which they were obtained. Note that although the pattern step from $x^{(3)}$ results in a failure the following exploratory move locates an improved point $x^{(4)}$ enabling a further pattern move to be made. The pattern and exploratory moves from $x^{(5)}$ however fail to produce a better point so the search recommences with an exploration from $x^{(5)}$ resulting in a new pattern direction.

The pattern search method is clearly a simple strategy which is easily programmed and which requires very little computer storage, and it has been found to be extremely useful in a wide variety of applications ranging from curve fitting to the on-line performance optimization of chemical processes (where its compactness can be of particular value).

It is of course very easy to modify the method to suit a particular application, which is especially useful if a given type of problem is to be solved a large number of times, as for instance in an on-line process optimization project. For example, one possible modification would be to carry out a linear search along each pattern direction. Another variation, called the Spider method, was proposed by Emery and O'Hagan (1966) who suggest that the steps in the exploratory moves are made, not along the co-ordinate directions, but along orthogonal directions whose orientation is chosen at random each time. This tactic is an attempt to prevent the search from terminating prematurely by foundering in a sharp valley, and the method is reported to have been used successfully on a number of microwave network optimization applications.

2.6 Razor search

Another modification of pattern search is that due to Bandler and Mcdonald (1969) who term their method *razor search* since it was devised specifically for problems in which the objective function exhibits a "razor sharp" valley, i.e. a valley along which a path of discontinuous derivatives lies.

As in the basic Hooke and Jeeves procedure the search generates a sequence of base points using exploratory and pattern moves. However, the step length used in the exploration is not held constant but depends upon the distance between the previous two base points; it thus expands and contracts according to the progress being made by the search. A second difference from the basic strategy is that whenever a pattern move followed by an exploratory move fails to improve the function, the pattern direction is not immediately rejected. First a reduced pattern step is taken,

$$\bar{x} = x^{(k)} + \alpha(x^{(k)} - x^{(k-1)}), \quad 0 < \alpha < 1,$$

followed by a new exploration with reduced step and if this proves successful the search proceeds as usual. If it also fails then a negative reduced pattern step is tried,

$$\bar{x} = x^{(k)} - \alpha(x^{(k)} - x^{(k-1)}),$$

followed by an exploration and only if this fails too does the search discard the pattern direction and perform a new exploration about the current base point in an attempt to define a more useful pattern.

When the search terminates a random move is made to a neighbouring point and a new search undertaken. If this second search results in the same solution then convergence is assumed, but if a different point is obtained then it is assumed that the direction defined by the two terminal points lies along a razor sharp valley and the search recommences with a pattern move along that direction. This procedure is repeated until two consecutive searches terminate at the same point.

2.7 Rosenbrock's Method

The method due to Rosenbrock (1960) also proceeds to the minimum by means of explorations along a set of directions, but defines at each iteration not just a single new direction but a complete new set.

For the kth iteration of the search, an approximation $x^{(k)}$ to the solution, a set of n mutually orthonormal search directions $p_1^{(k)}, \ldots, p_n^{(k)}$ and a set of n associated step lengths $\delta_1, \ldots, \delta_n$ are required. The co-ordinate directions provide a convenient initial choice for the search vectors.

The iteration begins by exploring the given set of directions in the following manner. Starting from $x^{(k)}$ a step δ_1 is taken along the direction $p_1^{(k)}$ and if it does not result in an increase in the value of the function it is considered to be successful, the new (improved) estimate for the minimum is retained, and δ_1 is multiplied by $\alpha > 1$. If the step is a failure it is rejected and δ_1 is multiplied by β where $0 > \beta > -1$. The search then considers $p_2^{(k)}$ in the same way, then $p_3^{(k)}$ and so on until all n directions have been explored whereupon it returns to $p_1^{(k)}$ again and continues to cycle around the directions in this manner until a success followed by a failure has been recorded at some time during the iteration for every direction. The effect of multiplying the step lengths by the factors α and β is to expand or to contract and reverse the exploration step in a direction according to whether the previous step in that direction succeeded or failed. The inclusion of equality in the definition of a successful step ensures that a success can always be obtained in a direction since after repeated failures the step length will become so small that, because of the finite computational accuracy of the computer, it causes no change in the function value.

The second part of the iteration is concerned with defining a new set of orthonormal search directions for use in the next iteration. If $x^{(k+1)}$ is the point arrived at as a result of the exploration from $x^{(k)}$ then the direction of total progress in the iteration is given by

$$s_1 = x^{(k+1)} - x^{(k)}$$

which can be written

$$s_1 = \sum_{j=1}^{n} \Delta_j p_j^{(k)} \tag{2.7.1}$$

where Δ_j is the algebraic sum of all the successful steps along the direction $p_j^{(k)}$. Since the exploration continues until at least one success has been achieved in each direction, $\Delta_j \neq 0$ for all j. The vector s_j where

$$s_j = s_{j-1} - \Delta_{j-1} p_{j-1}^{(k)}, \quad j = 2, \ldots, n \tag{2.7.2}$$

represents the total progress made in all directions other than $p_1^{(k)}, \ldots, p_{j-1}^{(k)}$. The new first direction for the next iteration is chosen to be the normalized direction of total progress in the present iteration, i.e.

$$p_1^{(k+1)} = \frac{s_1}{\|s_1\|},$$

and the remaining directions $p_j^{(k+1)}$, $j = 2, \ldots, n$, are chosen to form an orthonormal set with $p_1^{(k+1)}$ using the vectors s_j, $j = 2, \ldots, n$, and the Gram-Schmidt process. Thus

$$\left.\begin{aligned} d_j &= s_j - \sum_{i=1}^{j-1} (s_j^T p_i^{(k+1)}) p_i^{(k+1)} \\ p_j^{(k+1)} &= \frac{d_j}{\|d_j\|} \end{aligned}\right\} j = 2, 3, \ldots, n.$$

This completes the iteration and the next one begins from $x^{(k+1)}$ with explorations along the newly computed directions. Rosenbrock did not include any convergence test in his method and merely terminated the search after a specified number of function evaluations. However, he did suggest monitoring the value of $\|s_2\|/\|s_1\|$, which measures the progress made along all directions other than the first as a fraction of the total progress, and that if this remains generally less than 0·3 then it is unlikely that the minimum has been located.

The aim of the repeated orthonormalization is to align the first direction p_1 along the principal axis of the contours, p_2 along the best direction which can be found normal to p_1 and so on, and this prevents the oscillatory behaviour produced by the alternating variable method. As a result of his experience with the method Rosenbrock recommends values for the step adjustment parameters of $\alpha = 3$, $\beta = -0{\cdot}5$.

2.8 The D.S.C. method

Davies, Swann and Campey have suggested a procedure (Swann, 1964) similar to that of Rosenbrock in that it uses throughout a set of orthonormal directions which are rotated after each iteration, but which adopts a different exploration strategy.

Once again an iteration of the search requires an approximation $x^{(k)}$ to the minimum and a set of n orthonormal directions $p_1^{(k)}, \ldots, p_n^{(k)}$, again usually chosen initially as the co-ordinate directions, and this time just a single step length parameter δ. Instead of the stepping exploration used by

Rosenbrock a linear search is performed to locate the minimum along each direction in turn. The particular form of linear search proposed by Davies *et al.* is to take steps of increasing multiples of δ along $p_i^{(k)}$ in the direction of decreasing F until a bracket on the minimum is obtained whereupon a single quadratic interpolation is made to predict the position of the minimum more closely. Whichever of the interpolated point and the best of the points available before the interpolation corresponds to the smaller value of F is chosen to be the minimum in the direction $p_i^{(k)}$ and is used as the starting point for the search along $p_{i+1}^{(k)}$ (or $p_1^{(k+1)}$ if $i = n$).

When each of the current directions has been explored once in this way new search vectors are generated for the next iteration. Once again the new first direction $p_1^{(k+1)}$ is chosen to be the normalized direction of total progress in iteration k and the remaining directions are defined using the Gram–Schmidt relationships. However Δ_j in equations (2.7.1) and (2.7.2) now represents the total distance moved during the linear search along $p_j^{(k)}$ and because of the form of this search it is possible for any of the $\Delta_j, j = 1, \ldots, n$, to be zero. This would result in a breakdown in the orthonormalization process since at least one of the new directions $p_l^{(k+1)}$, $l = j, \ldots, n$, would be undetermined. The Rosenbrock method avoids this difficulty by requiring at least one successful step in each direction before terminating the exploration. In the D.S.C. method the problem is overcome by retaining for the next iteration any directions for which $\Delta_j = 0$ and omitting them from the orthonormalization procedure. Recently Powell (1968) and Palmer (1969) have suggested more economical methods for calculating the new search vectors which can be applied to both the method of Rosenbrock and the D.S.C. algorithm.

At the end of each iteration the total distance moved during the iteration, given by $\|s_1\|$, is compared with the current step length δ and whenever $\|s_1\| < \delta$ then δ is reduced; the search is terminated when δ becomes less than some pre-set limit.

Figure 2.5 shows the performance of the procedure on a function of two variables.

2.9 Evolutionary operation

A completely different approach to the problem is that which explores the parameter space by means of some geometric configuration of points rather than a set of directions. This type of method has developed from the technique of *evolutionary operation* (EVOP) originally devised by Box (1957).

EVOP was developed for the empirical optimization of plant performance in the presence of error. It is based on factorial designs and in its simplest two-level form it involves the determination of some measure of the plant

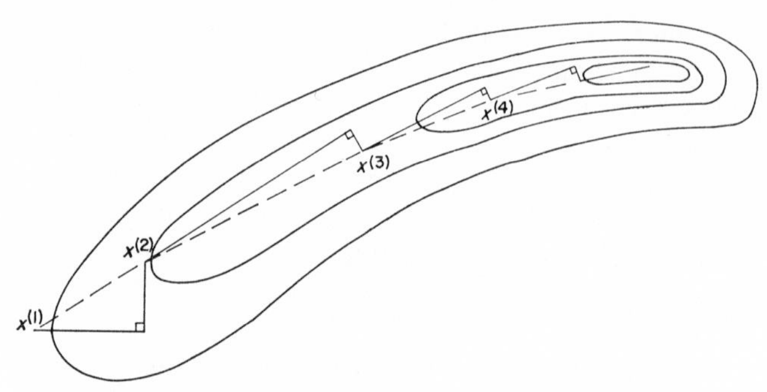

FIG. 2.5. The D.S.C. method for a function of two variables.

performance at each vertex and at the centre of a hypercube in the space of the independent variables of the plant. Replication is used to reduce the possibility of being misled by experimental or measuring errors, and then a new design is set up about whichever point of the vertices and centre corresponds to the best plant performance. If the centre is the best point then a contracted design is set up about it, but if the best point is a vertex then a new full-size design is set up with it as the centre. Repetitive use of this strategy forms the search procedure which, although devised for on-line use with industrial processes, is clearly readily applicable to situations in which the objective function values are obtained by evaluating some deterministic function.

However, although the procedure is very easily implemented on a computer, it is clearly expensive in terms of function evaluations, requiring 2^n+1 per design. The efficiency of the search can be improved by the use of fractional factorial designs in which the function is calculated at only 2^{n-m} of the vertices where $m \geqslant 1$, but it still remains unsatisfactory. A further possible improvement would be to use the relative magnitudes of the function at the vertices of the design to obtain an estimate for a useful direction to explore using a linear search routine, but this would still involve at least fractional designs at frequent intervals.

2.10 The simplex method

The first-order design requiring the smallest number of points is the regular simplex and Spendley, Hext and Himsworth (1962) suggested that a search strategy using such designs would be more efficient than one based on factorial designs. A regular simplex in n dimensions is $n+1$ mutually equidistant points so that for $n = 2$ it forms an equilateral triangle, for $n = 3$ a regular tetrahedron and so on, and it possesses the very useful property that a new simplex can be formed on any face of a given simplex by the addition of only a single new point.

The method begins by setting up a regular simplex in the space of the independent variables and evaluating the function at each vertex. If $x_0^{(1)}$ is a given initial estimate for the minimum, then a simplex of unit edge can be constructed by choosing the vertices to be $x_0^{(1)}, x_1^{(1)}, \ldots, x_n^{(1)}$ where

$$x_0^{(1)} = (x_{01}^{(1)}, x_{02}^{(1)}, \ldots, x_{0n}^{(1)})^T,$$

$$x_i^{(1)} = (x_{01}^{(1)}+\delta_1, x_{02}^{(1)}+\delta_1, \ldots, x_{0i-1}^{(1)}+\delta_1, x_{0i}^{(1)}+\delta_2, x_{0i+1}^{(1)}+\delta_1, \ldots, x_{0n}^{(1)}+\delta_1)^T, \quad i = 1, \ldots, n.$$

and

$$\delta_1 = \frac{(n+1)^{\frac{1}{2}}+n-1}{n\sqrt{2}},$$

$$\delta_2 = \frac{(n+1)^{\frac{1}{2}}-1}{n\sqrt{2}}.$$

The kth iteration of the search consists simply of replacing one of the vertices, say $x_j^{(k)}$, by its mirror image $x_j^{(k+1)}$ in the centroid of the remaining n vertices, i.e.

$$x_j^{(k+1)} = \frac{2}{n}(x_0^{(k)}+x_1^{(k)}+\ldots.+x_{j-1}^{(k)}+x_{j+1}^{(k)}+\ldots..+x_n^{(k)})-x_j^{(k)}.$$

Thus moving from one design to another costs only one function evaluation compared with the 2^n required when using two-level factorial designs. The point to be discarded, $x_j^{(k)}$, is normally chosen to be that vertex corresponding to the highest function value, which is considered to be the worst vertex of the current simplex. However, if $x_j^{(k-1)}$ was the worst vertex of the simplex at iteration $k-1$, then reflecting $x_j^{(k)}$ will produce a point $x_j^{(k+1)}$ which coincides with $x_j^{(k-1)}$ and the search will merely oscillate between the last two simplexes, so whenever the worst vertex is also the one most recently introduced the next to worst one is reflected instead.

The performance of the search on a function of two variables is illustrated in Fig. 2.6 in which the numbers indicate the order in which the vertices were introduced.

When one of the vertices is positioned in the neighbourhood of the minimum it will remain a vertex of the simplex and the search will rotate about it, attempts at further progress being frustrated by the size of the simplex as can be seen in Fig. 2.6 in which vertex 15 is placed close to the minimum. From their experience with the method Spendley *et al* deduced that the maximum expected age of any vertex could be approximately represented by

$$\alpha = 1{\cdot}65n + 0{\cdot}05n^2$$

so that if a vertex exceeds this age it is reasonable to conclude that this persistent vertex is in the neighbourhood of the minimum. On such an occurrence the simplex is contracted by reducing the distance of each vertex

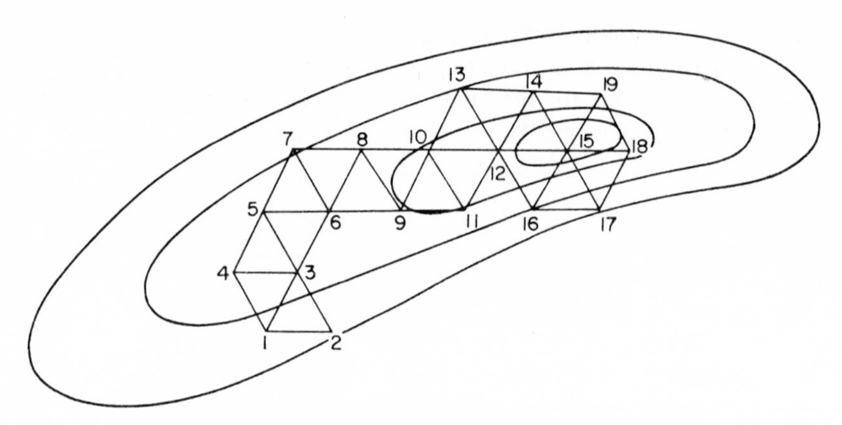

FIG. 2.6. The simplex method for a function of two variables.

from the oldest one and the search recommences using the new, smaller, design. The search ends when the size of the simplex falls below some preassigned limit.

2.11 Simplex method of Nelder and Mead

The *simplex method* provides a useful procedure for minimizing a given function and a number of variations of it have been produced. Probably the most effective version is that due to Nelder and Mead (1965) in which the regularity of the design is abandoned and the simplex automatically rescales itself according to the local geometry of the function under investigation.

Suppose that for iteration k the vertices of the simplex are $x_0^{(k)}, x_1^{(k)}, \ldots, x_n^{(k)}$ with corresponding function values $F_0, F_1, \ldots, F_n$ ordered such that

$$F_n > F_{n-1} > \ldots > F_1 > F_0.$$

Hence $x_0^{(k)}$ is the best vertex of the simplex and $x_n^{(k)}$ is the worst. Let $c^{(k)}$ be the centroid of the vertices $x_0^{(k)}, x_1^{(k)}, \ldots, x_{n-1}^{(k)}$, i.e.

$$c_i^{(k)} = \frac{1}{n}\sum_{j=0}^{n-1} x_{ji}^{(k)}, \quad i = 1, \ldots, n.$$

Then as in the original simplex method the worst vertex $x_n^{(k)}$ is to be replaced and a simple reflection move is tried first, giving a new point $x_r^{(k)}$ where

$$x_r^{(k)} = c^{(k)} + \alpha(c^{(k)} - x_n^{(k)})$$

and $\alpha > 0$ is the reflection coefficient. There are then three possible cases to be considered: $x_r^{(k)}$ is a point such that $F_0 < F_r < F_{n-1}$; $F_r < F_0$ so that $x_r^{(k)}$ would be a new best point; $F_r > F_{n-1}$ so that $x_r^{(k)}$ would be a new worst point.

In the case where $F_0 < F_r < F_{n-1}$ then $x_r^{(k)}$ replaces $x_n^{(k)}$ and the iteration is complete.

However, if the reflection has produced a new best point, then the direction of the reflection may be a useful one so an attempt is made to expand the design along it by defining the point

$$x_e^{(k)} = c^{(k)} + \beta(x_r^{(k)} - c^{(k)})$$

where $\beta > 1$ is the expansion coefficient. Then if $F_e < F_0$ the expansion is considered to be successful and $x_e^{(k)}$ replaces $x_n^{(k)}$; otherwise the expansion is deemed to have failed and $x_n^{(k)}$ is replaced by $x_r^{(k)}$. In either event the iteration is then complete.

If the original reflection resulted in a new worst point then it is assumed that the size of the design is too large to allow any progress to be made, and so a contracted simplex is derived by defining the point $x_c^{(k)}$ where

$$x_c^{(k)} = c^{(k)} + \gamma(x_n^{(k)} - c^{(k)}) \quad \text{if } F_n < F_r$$

and

$$x_c^{(k)} = c^{(k)} + \gamma(x_r^{(k)} - c^{(k)}) \quad \text{if } F_n > F_r,$$

and γ is the contraction coefficient with $0 < \gamma < 1$. If $F_c < \min(F_n, F_r)$ the contraction has succeeded and $x_c^{(k)}$ replaces $x_n^{(k)}$; otherwise a more comprehensive contraction is carried out by halving the distances from the best point $x_0^{(k)}$ of all the other vertices of the simplex. In either case the iteration is then complete.

The reflection, expansion and contraction steps for a function of two variables are illustrated in Fig. 2.7.

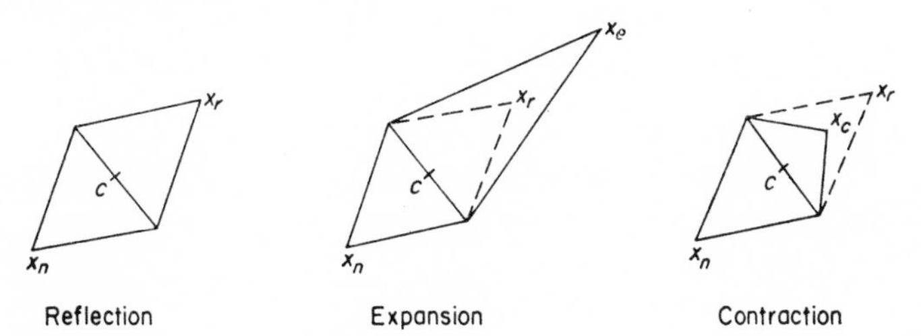

FIG. 2.7. Nelder and Mead reflection, expansion and contraction steps for a function of two variables.

The convergence criterion is based on the variation in the function values over the simplex and the search is concluded when the standard deviation σ of the function values at the vertices falls below some pre-assigned limit, where

$$\sigma = \sqrt{\sum_{i=0}^{n} \frac{(F_i - \bar{F})^2}{n}}$$

and $\bar{F}$ is the mean of the function values.

2.12 Conclusion

The primary reason for interest in direct search methods is their generality; they require only the ability to evaluate the objective function at specified points of E^n, and since they usually make no more assumptions other than continuity of the function they can be applied to a very wide class of problems. They are thus particularly useful for cases where the function is non-differentiable, where the first partial derivatives of the function are discontinuous, and where the function value depends upon some physical measurement and may therefore be subject to random error, all of which are problems which can cause difficulties using the more theoretically based gradient methods. Although the methods described above have been developed heuristically and no proofs of convergence have been derived for them, in practice they have generally proved to be robust and reliable in that only rarely do they fail to locate at least a local minimum of a given function, although sometimes the rate of convergence can be very slow.

In addition, the relative simplicity of the direct search approach compared with the gradient techniques described in later chapters can prove advantageous since it generally means that the methods can be easily and quickly programmed, are computationally compact and make only very modest demands on storage, all of which can be important considerations in practical optimization projects.

3. Problems Related to Unconstrained Optimization

M. J. D. Powell
Atomic Energy Research Establishment, Harwell

3.1 Introduction

In the other chapters in this volume, the main problem that is considered is the calculation of the least value of a given general function $F(x)$, where x is a vector of n real variables, and where there are no constraints on the values of the variables. However, in this chapter we will consider two related problems, that have much in common with the main theme of this volume. The first related problem is to calculate the least value of $F(x)$ in the special case when it is a sum of squares

$$F(x) = \|f(x)\|^2 = \sum_{t=1}^{m} [f_t(x)]^2, \quad m \geqslant n, \tag{3.1.1}$$

and the second related problem is to calculate the least value of a general function $F(x)$, given that there are constraints on the values of the variables, for example $\|x\|^2 \leqslant 1$. Some techniques for taking account of constraints will be described, that depend mainly on the availability of a good algorithm for unconstrained minimization.

General algorithms for unconstrained minimization can be applied to the function (3.1.1), but usually it is much more efficient to use an algorithm that takes account of the fact that $F(x)$ is a sum of squares, because then it is straightforward to estimate second derivatives from calculated first derivatives. This remark is explained in section 3.2, and then some algorithms are described for solving the least squares problem, that require the user to provide a subroutine that evaluates the first derivatives of the functions $f_t(x)$ $(t = 1, 2, \ldots, m)$. Bard's (1970) numerical examples indicate the gain in efficiency that can be obtained by using a "least squares algorithm" in place of a "general algorithm for unconstrained optimization".

In section 3.3 more "least squares algorithms" are described, but these are for the case when the user does not program the evaluation of any derivatives, so he provides a subroutine to calculate only the function values $f_t(x)$ $(t = 1, 2, \ldots, m)$ for any x.

Some of these algorithms are intended to solve a system of non-linear equations

$$f_t(x) = 0, \quad t = 1, 2, \ldots, n, \tag{3.1.2}$$

which, in the notation of equation (3.1.1), is the special case $m = n$.

If one has to calculate the least value of a function $F(x)$ and there are constraints on the variables, then sometimes it is possible to take account of the constraints by a change of variables $x = x(w)$, in which case the objective function becomes

$$F(x(w)) \equiv \bar{F}(w) \tag{3.1.3}$$

say. By a careful choice of the mapping $x(w)$ it can happen that the least value of $\bar{F}(w)$, subject to no constraints on the variables, is equal to the least value of $F(x)$ subject to the original constraints. This method is discussed in section 3.4. Of course, when it is used an algorithm for unconstrained minimization is applied to the function $\bar{F}(w)$.

Unfortunately it often happens that there are constraints on the variables, and it is not possible to find a change of variables $x = x(w)$ such that the method of section 3.4 can be used. However, most constrained problems can be converted to unconstrained problems by "penalty function methods", so these methods are considered in section 3.5. In a penalty function method a function ψ, depending on a parameter ρ, is added to $F(x)$, giving the function

$$\Phi(x, \rho) = F(x) + \psi(x, \rho). \tag{3.1.4}$$

Then an algorithm for unconstrained optimization is applied to the function $\Phi(x, \rho)$, for a sequence of positive values of ρ that converges to zero. By a careful choice of $\psi(x, \rho)$ it can be arranged that the required value of $F(x)$ is the limit of the minima of $\Phi(x, \rho)$ as ρ tends to zero. For example, $\psi(x, \rho)$ may satisfy the three conditions (i) $\psi(x, \rho) \geqslant 0$, (ii) If x satisfies the constraints then $\psi(x, \rho)$ tends to zero as ρ tends to zero, and (iii) If x violates the constraints then $\psi(x, \rho)$ tends to infinity as ρ tends to zero.

An easy way of satisfying these conditions is to let $\psi(x, \rho) = 0$, unless x violates the constraints when we let $\psi(x, \rho) = \infty$, but if we made this choice then the unconstrained minimization of $\Phi(x, \rho)$ would be practically impossible. Therefore the purpose of the parameter ρ is that $\Phi(x, \rho)$ is a smooth function for $\rho > 0$, but this smoothness has to become less and less real as ρ tends to zero.

The remarks of the last paragraph show a serious disadvantage of penalty function methods, which is that as ρ tends to zero the calculation of the least value of $\Phi(x, \rho)$ becomes awkward. Therefore it is worthwhile to replace the function $\psi(x, \rho)$ of equation (3.1.4) by another suitable function, which has the extra property that it is uniformly smooth for all values of x

that may occur during an unconstrained minimization calculation. Powell (1969a), for instance, suggests a suitable function, that depends on two vectors of parameters, l and h, whose values are adjusted automatically. For each choice of parameter values a method for unconstrained minimization is applied to calculate the least value of the function (3.6.6), and we let $x(l, h)$ be the resultant vector of variables. It can be proved that $x(l, h)$ converges to the solution of the constrained problem. Another promising method is due to Fletcher (1970a), and it is especially useful because it can solve many constrained minimization problems by a single unconstrained minimization, instead of by a sequence of calculations. These algorithms are described in section 3.6, and we call them Lagrange parameter methods, because they avoid the unbounded functions of the usual penalty function methods, by strategies that are similar to the classical method of Lagrange parameters for treating equality constraints.

3.2 Least squares algorithms requiring first derivatives

We wish to calculate the least value of the sum of squares

$$F(x) = \|f(x)\|^2 = \sum_{t=1}^{m} [f_t(x)]^2, \quad m \geqslant n, \tag{3.2.1}$$

given that we have a computer subroutine that evaluates the functions $f_t(x)$ $(t = 1, 2, \ldots, m)$ and the first derivative matrix

$$J_{tj}(x) = \frac{\partial f_t(x)}{\partial x_j} \quad (t = 1, 2, \ldots, m;\ j = 1, 2, \ldots, n) \tag{3.2.2}$$

for any x. Algorithms for this calculation are iterative, and they calculate a sequence of points $x^{(1)}, x^{(2)}, \ldots$, that should converge to a point x^* that minimizes $F(x)$.

The algorithms for non-linear least squares problems calculate $x^{(k+1)}$ from linear approximations to the functions $f_t(x)$, these approximations being chosen so that they are good near $x^{(k)}$. Specifically on the kth iteration the function $f_t(x)$ is approximated by the linear function

$$l_t^{(k)}(x) = f_t(x^{(k)}) + \sum_{j=1}^{n} J_{tj}^{(k)}\{x_j - x_j^{(k)}\}, \quad t = 1, 2, \ldots, m, \tag{3.2.3}$$

where the notation $J_{tj}^{(k)}$ means the first derivative $J_{tj}(x^{(k)})$. Note that if $f_t(x)$ is twice differentiable at $x^{(k)}$, then the error $\{l_t^{(k)}(x) - f_t(x)\}$ is of order $\|x - x^{(k)}\|^2$. The approximations (3.2.3) lead to the approximation

$$F(x) \approx \sum_{t=1}^{m} [l_t^{(k)}(x)]^2 = \Lambda^{(k)}(x) \tag{3.2.4}$$

say.

In section 3.1 we mentioned that the first derivatives (3.2.2) provide estimates of the second derivatives of $F(x)$, and these estimates are the second derivatives of $\Lambda^{(k)}(x)$,

$$\frac{\partial^2 F(x)}{\partial x_i \, \partial x_j} \approx \frac{\partial^2 \Lambda^{(k)}(x)}{\partial x_i \, \partial x_j},$$
$$= 2 \sum_{t=1}^{m} J_{ti}^{(k)} J_{tj}^{(k)},$$
$$= B_{ij}^{(k)} \tag{3.2.5}$$

say. This equation shows that $B^{(k)}$ is the matrix $2J^{(k)T}J^{(k)}$, and therefore it is positive definite or positive semidefinite. It follows that it is usually straightforward to calculate the least value of $\Lambda^{(k)}(x)$, and in the case when $B^{(k)}$ is positive definite this least value is obtained when x is the vector $[x^{(k)} - 2\{B^{(k)}\}^{-1}\{J^{(k)T}f(x^{(k)})\}]$. Therefore the classical "Gauss–Newton method" uses the iteration

$$x^{(k+1)} = x^{(k)} - 2\{B^{(k)}\}^{-1}\{J^{(k)T}f(x^{(k)})\}, \quad k = 1, 2, \ldots. \tag{3.2.6}$$

Whether or not it converges depends of course on the goodness of the approximations (3.2.3), and also it depends in a remarkable way on some other factors, including the function values $f_t(x^*)$ $(t = 1, 2, \ldots, m)$. We now discuss the convergence of the Gauss–Newton iteration.

First we consider a numerical example, in which $n = 1$ and $m = 2$: minimize the function

$$F(x) = (x+1)^2 + (\lambda x^2 + x - 1)^2, \tag{3.2.7}$$

where λ is a parameter. In this case the matrix $B^{(k)}$ is 1×1, and its element has the value $2\{1 + (2\lambda x^{(k)} + 1)^2\}$. Therefore the Gauss–Newton iteration is defined by the equation

$$x^{(k+1)} = \frac{2\lambda^2 x^{(k)3} + \lambda x^{(k)2} + 2\lambda x^{(k)}}{1 + (2\lambda x^{(k)} + 1)^2}, \tag{3.2.8}$$

and in particular if $\lambda = 0$, then $x^{(k+1)} = 0$, so we minimize expression (3.2.7) in one iteration, which is expected because in this case the approximations (3.2.3) are exact. However, if $\lambda \neq 0$ and $|x^{(k)}|$ is very small, then equation (3.2.8) gives the expression

$$x^{(k+1)} = \lambda x^{(k)} + 0\|x^{(k)}\|^2. \tag{3.2.9}$$

It follows that if the limit of the sequence $x^{(k)}$ $(k = 1, 2, \ldots)$ is zero, the rate of convergence is only linear, although the difference between $l_t^{(k)}(x)$ and $f_t(x)$ is of order $\|x - x^{(k)}\|^2$.

This linear rate of convergence is due to the error of the approximation (3.2.5), because to make an accurate prediction of the least value of a function, one usually requires an accurate estimate of the curvature of the function. In our example the second derivative of $F(x)$ at $x^{(k)}$ has the value $\{12\lambda^2 x^{(k)2} + 12\lambda x^{(k)} + 4 - 4\lambda\}$, so the ratio of the left-hand side to the right-hand side of the approximation (3.2.5) is $\{1 - \lambda + 0\|x^{(k)}\|\}$. Therefore we expect $\{x^{(k+1)}-x^{(k)}\}$ to differ from $\{x^*-x^{(k)}\}$ by about the factor $(1-\lambda)$, which confirms expression (3.2.9).

In the general case (3.2.1) we calculate the left-hand side of the approximation (3.2.5) at $x = x^{(k)}$, and we find that this approximation reduces to

$$2\sum_{t=1}^{m}\left\{J_{ti}^{(k)}J_{tj}^{(k)}+f_t(x^{(k)})\frac{\partial^2 f_t(x^{(k)})}{\partial x_i\,\partial x_j}\right\}\approx B_{ij}^{(k)}. \tag{3.2.10}$$

Therefore the error of the approximation is the term

$$2\,\Sigma f_t(x^{(k)})\,\partial^2 f_t(x^{(k)})/\partial x_i\,\partial x_j,$$

and it will be small if the functions $f_t(x)$ are nearly linear, or if the function values $f_t(x^{(k)})$ are small. Further, because the inverse of the matrix $B^{(k)}$ occurs in equation (3.2.6), we would like the error of this approximation to be small in comparison with the least eigenvalue of the matrix $B^{(k)}$. Fortunately it happens quite often in practice that the error of the approximation (3.2.10) is sufficiently small for the Gauss–Newton method to be useful. However, we cannot expect the final rate of convergence to be better than linear unless $F(x^*) = 0$.

It is interesting to consider the example (3.2.7) when $\lambda < -1$. In this case it is easy to verify that $F(x)$ is strictly convex, and that its minimum is at the point $x = 0$. But expression (3.2.9) shows that the Gauss–Newton iteration will not permit the sequence $x^{(k)}$, $k = 1, 2, \ldots$, to converge to the required limit. Therefore this algorithm is not suitable for all general least squares problems.

A useful modification to the Gauss–Newton iteration, to prevent divergence of the sequence $x^{(1)}, x^{(2)}, \ldots$, is to let the correction vector of equation (3.2.6) be a search direction in the space of the variables. Then the kth iteration of the algorithm sets

$$x^{(k+1)} = x^{(k)}+\alpha^{(k)}p^{(k)}, \tag{3.2.11}$$

where $\alpha^{(k)}$ is a scalar, and where $p^{(k)}$ is the vector

$$p^{(k)} = -2\{B^{(k)}\}^{-1}\{J^{(k)T}f(x^{(k)})\}. \tag{3.2.12}$$

If the value of $\alpha^{(k)}$ is calculated to minimize the function of one variable

$$\phi^{(k)}(\alpha) = F(x^{(k)}+\alpha p^{(k)}), \tag{3.2.13}$$

then Kowalik and Osborne (1968) call the iteration (3.2.11) the "modified Gauss algorithm". However, in practice it is not sensible to calculate $F(x)$ many times in order to fix the value of $\alpha^{(k)}$. Instead it is preferable to accept a coarse estimate of the value that minimizes the function (3.2.13). Further details are not given here because Murray discusses the minimization of a function of one variable in the first chapter of this book. Also some methods for approximate linear searches are described by Bard (1970) and by Maddison (1966).

If the matrix $B^{(k)}$ is singular or nearly singular, then the calculation of $p^{(k)}$ (see equation (3.2.12)) may be awkward. But because $B^{(k)} = 2J^{(k)T}J^{(k)}$, and because $J^{(k)}$ is an $m \times n$ matrix, it frequently happens that the eigenvalues of $B^{(k)}$ are bounded away from zero, especially when $m > n$. If, in addition, the matrices $B^{(k)}$ are bounded above, and if the set

$$\{x | F(x) \leqslant F(x^{(1)})\}$$

is bounded, then it can be proved that the modified Gauss algorithm makes the vectors $g^{(k)}$ $(k = 1, 2, \ldots)$ converge to zero, where $g^{(k)}$ is the first derivative vector of $F(x)$ at $x^{(k)}$. Therefore usually the sequence of vectors $x^{(k)}$ converges to a minimum of $F(x)$ (Hartley, 1961).

It is not difficult to prove this convergence theorem, because our assumptions and equation (3.2.12) show that the search direction $p^{(k)}$ is equal to $g^{(k)}$ multiplied by a negative-definite symmetric matrix whose condition number is bounded. It follows that, if instead of calculating $\alpha^{(k)}$ to minimize the function (3.2.13), we prefer to use few evaluations of $F(x)$ to fix $\alpha^{(k)}$, then the convergence theory published by Wolfe (1969) is applicable.

However, if the matrices $B^{(k)}$ are not bounded away from singularity, then it can happen that the modified Gauss algorithm generates a sequence of vectors $x^{(k)}$ $(k = 1, 2, \ldots)$, that converges very quickly to a point that is not even a stationary point of $F(x)$ (Powell, 1970a).

Usually the modified Gauss algorithm works well, but sometimes the iterations make slow progress, and in these cases it frequently happens that the search directions $p^{(k)}$ are nearly at right angles to the gradients $g^{(k)}$. Therefore Marquardt (1963) suggests an algorithm in which the direction of $p^{(k)}$ is biased towards the steepest-descent direction. The idea behind this algorithm was first proposed by Levenberg (1944). It is to let $p^{(k)}$ be the vector

$$p^{(k)} = -2\{B^{(k)} + \lambda^{(k)} I\}^{-1}\{J^{(k)T} f(x^{(k)})\}, \tag{3.2.14}$$

for some non-negative value of $\lambda^{(k)}$. We note that if $\lambda^{(k)} = 0$, we have the Gauss direction (3.2.12), and if $\lambda^{(k)}$ is made large then $p^{(k)}$ tends to $-g^{(k)}/\lambda^{(k)}$ (where $g^{(k)}$ is the gradient of $F(x)$ at $x^{(k)}$), so the search direction tends to be a multiple of the steepest-descent vector.

Because in Marquardt's method the value of $\lambda^{(k)}$ can be used to control the length of $p^{(k)}$, one may always set $\alpha^{(k)} = 1$ in equation (3.2.11). However, if the value of $\lambda^{(k)}$ is changed, then one has to solve a new set of linear equations to calculate the new $p^{(k)}$, so it may be more economical to obtain $F(x^{(k+1)}) < F(x^{(k)})$ by working on the function (3.2.13). Then if one needs to choose $\alpha^{(k)} < 1$, one sets $\lambda^{(k+1)} > \lambda^{(k)}$.

We have noted that $\lambda^{(k)}$ in equation (3.2.14) introduces a bias towards the steepest-descent direction, and therefore it can be proved that if the first derivative vector of $F(x)$ is a uniformly continuous function of x for

$\{x|F(x) \leqslant F(x^{(1)})\}$, then one can choose $\lambda^{(k)}$ $(k = 1, 2, \ldots)$ so that $g^{(k)}$ converges to zero. Here we do not need the condition that the set $\{x|F(x) \leqslant F(x^{(1)})\}$ is bounded. It was needed in the theorem on the modified Gauss algorithm to ensure that a finite value of α minimizes the function (3.2.13), but in Marquardt's method we do not attempt to minimize this function. Note that Marquardt's method does not break down on the function (Powell, 1970a) that causes the modified Gauss algorithm to converge to a point that is not a stationary point of $F(x)$.

An interesting interpretation of the role of $\lambda^{(k)}$ in equation (3.2.14) comes from the observation that the vector (3.2.14) minimizes the quadratic function of p

$$\sum_{t=1}^{m} [l_t^{(k)}(x^{(k)}+p)]^2 + \tfrac{1}{2}\lambda^{(k)}\|p\|^2. \tag{3.2.15}$$

Therefore the vector $x^{(k)}+p^{(k)}$ minimizes the function $\Lambda^{(k)}(x)$ {see equation (3.2.4) for the definition of $\Lambda^{(k)}(x)$}, subject to a restriction on the length of $p^{(k)}$. Consequently it can be proved that if d is any vector satisfying the conditions $d \neq p^{(k)}$ and $\|d\| \leqslant \|p^{(k)}\|$, then the inequality

$$\Lambda^{(k)}(x^{(k)}+d) > \Lambda^{(k)}(x^{(k)}+p^{(k)})$$

holds (Marquardt, 1963).

A comparison of methods for minimizing a sum of squares, in the case that first derivatives are calculated, is made by Bard (1970). Unfortunately none of his least squares methods are standard, for he modifies the "modified Gauss algorithm" in three ways, and also he changes Marquardt's algorithm. His numerical results indicate that a modified Gauss algorithm is marginally better, but I would prefer Marquardt's algorithm because the theory indicates that it is more reliable.

Bard also includes in his comparison some general minimization algorithms, that do not take advantage of the fact that $F(x)$ is a sum of squares. His examples show that these algorithms are not as efficient as the least squares methods, so, in spite of the fact that we have noted that the ultimate convergence rate of a least squares algorithm is usually only linear, in general it seems that one should prefer a special method if $F(x)$ is a sum of squares. If, however, one uses a general method that requires an initial estimate of the second derivative matrix of $F(x)$, then it is usually advantageous to let this matrix be the matrix $B^{(1)}$ of equation (3.2.5).

The main disadvantage in practice of the least squares gradient methods is that the amount of computing per iteration is of order (mn^2+n^3), due to the need to solve a system of linear equations when applying formula (3.2.6), (3.2.12) or (3.2.14). This fact is tolerable if n is small, or if the solution of the linear equations requires much less computing than the calculation of the derivatives (3.2.2), but occasionally most of the work of using a least squares algorithm is dominated by the calculation of $p^{(k)}$ $(k = 1, 2, \ldots)$. In this case it may be better to use a general minimization algorithm that requires only of order mn computing operations per iteration, or it may be better to use one of the algorithms described in Section 3.3 of this paper. Another useful way of reducing the amount of computation per iteration is to keep one matrix $B^{(k)}$ for several iterations. For example, one could let $B^{(k)} = B^{(k-1)}$ unless $k = 5j+1$, for some integer j, when one would define $B^{(k)}$ by the usual formula (3.2.5). Thus only every fifth iteration would require of order (mn^2+n^3) operations, and the remaining iterations would require of order mn computer operations.

Further, if the matrix $B^{(k)}$ is held constant for five iterations, then it may be sensible to change the matrix $J^{(k)T}$ of equations (3.2.12) and (3.2.14) only on every fifth iteration. However, this is a more drastic modification, because if the matrix $J^{(k)}$ is wrong and $x^{(k)}$ is not a stationary point of $F(x)$, then using Marquardt's algorithm it may not be possible to obtain the inequality $F(x^{(k+1)}) < F(x^{(k)})$. However, if the matrices $J^{(k)}$ are correct and the matrices $B^{(k)}$ are wrong, then usually Marquardt's algorithm finds a minimum of $F(x)$. Nevertheless the idea of keeping $J^{(k)}$ constant for a few iterations can be useful when the calculation of derivatives is much more expensive than the calculation of function values.

For further reading on gradient methods for minimizing a sum of squares, the book by Ortega and Rheinboldt (1970) is recommended. Also a new algorithm has been published recently by Jones (1970).

3.3 Least squares algorithms that use only function values

The algorithms of Section 3.2 require a computer subroutine to calculate the first derivatives (3.2.2), but for obvious reasons computer users prefer not to calculate any derivatives. Therefore some algorithms have been proposed for minimizing a sum of squares, using only values of the functions $f_t(x)$ $(t = 1, 2, \ldots, m)$ in their search for the least value of $F(x)$. They are discussed in this section.

The algorithms are based on the methods that are used when derivatives are available, so the main new feature is the problem of estimating first derivatives. One method of estimation is to use difference approximations of the form

$$\frac{\partial f_t(x)}{\partial x_j} \approx \frac{f_t(x+\eta e_j)-f_t(x)}{\eta}, \tag{3.3.1}$$

where e_j is the jth column of the unit matrix, and where η is a small positive scalar, but there are three disadvantages in this procedure. The first is the common numerical difficulty of choosing an adequate value of η. The second disadvantage is that to estimate a first derivative matrix the function values

$f_t(x)$ $(t = 1, 2, \ldots, m)$ have to be calculated at $n+1$ different points x, so most function values may be used to estimate derivatives, instead of being used directly in the main calculation to minimize $F(x)$. Thirdly, if η is so small that the estimate (3.3.1) is quite accurate, then the points x, at which the function values $f_t(x)$ are calculated, tend to group into bunches of $n+1$ points, which is an unsatisfactory disposition of function evaluations. Therefore in the methods that are described in this section, we prefer not to use the approximation (3.3.1), except that before the first iteration it is common to take steps along the co-ordinate directions, in order to obtain an initial estimate of the first derivative matrix.

It is my opinion that the disadvantages of difference approximations like expression (3.3.1) become stronger if n is made larger. However, the case $m = n = 1$ demonstrates the inefficiency of difference approximations. In this case if each iteration uses equation (3.2.6), the derivative being estimated by setting a very small value of η in expression (3.3.1), then two function evaluations are required per step, and the process is like the well-known Newton–Raphson iteration, so the mean rate of convergence per function evaluation is 1·414. If, however, the secant method is used instead, then there is only one function evaluation per iteration, and the rate of convergence is 1·618 (Jarratt, 1970).

It is straightforward to generalize the secant method for solving one equation in one unknown to the problem of minimizing the function (3.2.1) (Wolfe, 1959). For example, the following iterative algorithm suggests itself. Given $x^{(k-n)}, x^{(k-n+1)}, \ldots, x^{(k)}$ and $f(x^{(k-n)}), f(x^{(k-n+1)}), \ldots, f(x^{(k)})$, calculate the linear functions $l_t^{(k)}(x)$ $(t = 1, 2, \ldots, m)$ that satisfy the equations

$$l_t^{(k)}(x^{(j)}) = f_t(x^{(j)}), \quad j = k-n, k-n+1, \ldots, k, \tag{3.3.2}$$

and then let $x^{(k+1)}$ be the vector of variables that minimizes the quadratic function (3.2.4). Here we are using equation (3.2.6), except that the first derivative matrix $J^{(k)}$ is obtained by fitting the changes in $f(x)$ that have been made by the most recent n iterations. To begin the iterations the vector $x^{(1)}$ has to be given by the user of the algorithm, and then, for $j = 1, 2, \ldots, n$, we let $x^{(j+1)}$ equal $x^{(j)}$ plus a small step along the jth co-ordinate direction.

Unfortunately this algorithm retains the disadvantages of the Gauss–Newton method, and also it has another serious defect, which is now demonstrated by the numerical example: minimize the function

$$F(x) = (x_1+x_2-4)^2+(x_2^2-6x_2+8)^2. \tag{3.3.3}$$

Let us start the calculation at the point $x^{(1)} = (0, 0)^T$, and let $x^{(2)} = (0·1, 0)^T$ and $x^{(3)} = (0·1, 0·1)^T$. Then the next three iterations give the numbers

$$J^{(3)} = \begin{pmatrix} 1 & 1 \\ 0 & -5·9 \end{pmatrix} \qquad x^{(4)} = \begin{pmatrix} 2·6441 \\ 1·3559 \end{pmatrix} \quad f^{(4)} = \begin{pmatrix} 0 \\ 1·7030 \end{pmatrix}$$

$$J^{(4)} = \begin{pmatrix} 1 & 1 \\ 0.6694 & -5.9 \end{pmatrix} \qquad x^{(5)} = \begin{pmatrix} 2.3849 \\ 1.6151 \end{pmatrix} \qquad f^{(5)} = \begin{pmatrix} 0 \\ 0.9179 \end{pmatrix}$$

$$J^{(5)} = \begin{pmatrix} 1 & 1 \\ -0.5008 & -3.5297 \end{pmatrix} \quad x^{(6)} = \begin{pmatrix} 2.0819 \\ 1.9181 \end{pmatrix} \qquad f^{(6)} = \begin{pmatrix} 0 \\ 0.1704 \end{pmatrix}$$

We note that $F(x)$ is decreasing well. However, it is impossible to define $J^{(6)}$ from the vectors $\{x^{(j)}, f(x^{(j)}), j = 4, 5, 6\}$, because the points $x^{(4)}$, $x^{(5)}$ and $x^{(6)}$ are collinear.

The "generalized secant method" breaks down in the above example because the function $f_1(x)$ is exactly linear, so we conclude that if we use equation (3.3.2) to define the linear approximation $l_t^{(k)}(x)$ to the function $f_t(x)$, then it is necessary to ensure that the points $x^{(j)}$ are linearly independent.

Spendley (1969) and Peckham (1970) suggest using more than $(n+1)$ function values $f(x)$ to define $l^{(k)}(x)$, and they try to satisfy equation (3.3.2) for $j = k-r, k-r+1, \ldots, k$, where r is an integer in the range $n \leqslant r \leqslant 3n$. They do this because it may happen that, although the points $x^{(k-n)}$, $x^{(k-n+1)}, \ldots, x^{(k)}$ are dependent, the points $x^{(k-r)}, x^{(k-r+1)}, \ldots, x^{(k)}$ are linearly independent, so letting r exceed n can avoid linear dependence. If r exceeds n then there are more equations than unknowns to define the function $l_t^{(k)}(x)$, so they solve the equations in a least squares sense. However, this device is not always satisfactory because, for the example (3.3.3), the points $x^{(4)}, x^{(5)}, \ldots, x^{(k)}$ are collinear for all $k \geqslant 4$. Peckham (1970) is aware of the difficulty shown by example (3.3.3), and he uses a "pseudo-random number procedure" if he finds any ill-conditioning of linear equations.

In practice, if one uses the generalized secant method, one should not actually solve the equations (3.3.2), because this work is unnecessary. Instead we define the vectors

$$\left.\begin{aligned} s^{(j)} &= x^{(j+1)} - x^{(j)} \\ d^{(j)} &= f(x^{(j+1)}) - f(x^{(j)}) \end{aligned}\right\}, \tag{3.3.4}$$

and we note that the equations (3.3.2) imply the identity

$$\begin{aligned} l^{(k)}\left(x^{(k)} + \sum_{i=1}^{n} z_i s^{(k-n+i-1)}\right) &= f(x^{(k)}) + \sum_{i=1}^{n} z_i d^{(k-n+i-1)} \\ &= f(x^{(k)}) + D^{(k)} z \end{aligned} \tag{3.3.5}$$

for all vectors z, where $d^{(k-n+i-1)}$ is the ith column of the matrix $D^{(k)}$. Therefore the vector z that minimizes the sum of squares (3.2.4) is the vector

$$z^{(k)} = -\{D^{(k)T} D^{(k)}\}^{-1} D^{(k)T} f(x^{(k)}), \tag{3.3.6}$$

so we may replace equation (3.2.6) by the equivalent definition

$$x^{(k+1)} = x^{(k)} + \sum_{i=1}^{n} z_i^{(k)} s^{(k-n+i-1)}. \tag{3.3.7}$$

It follows that on each iteration the only part of the calculation that may require of order mn^2 computer operations is the calculation of the vector (3.3.6). Even this calculation can be reduced to of order mn operations, by storing and updating the matrices $\{D^{(k)T}D^{(k)}\}^{-1}$ ($k = 1, 2, \ldots$), using the method described by Rosen (1960), because the matrices $\{D^{(k)T}D^{(k)}\}$ and $\{D^{(k+1)T}D^{(k+1)}\}$ differ in only one row and column for each integer k.

The method described in the last paragraph is the basis of Powell's (1965) least squares algorithm. However, this algorithm includes two modifications of the generalized secant method. The first is that a vector corresponding to $\sum z_i^{(k)} s^{(k-n+i-1)}$ (see equation (3.3.7)) is used as a *search* direction, analogous to $p^{(k)}$ in equation (3.2.11), and the second modification is that to estimate first derivatives the vectors $s^{(j)}$ and $d^{(j)}$ ($j = k-n, k-n+1, \ldots, k-1$) may not be used. Instead one or more pairs ($s^{(j)}, d^{(j)}$), where $j < k-n$, may take the place of more recently calculated vectors, in order to avoid the linear dependence shown by the example (3.3.3). Specifically for the $(k+1)$th iteration the pair $(s^{(k)}, d^{(k)})$ replaces just one of the pairs $(s^{(j)}, d^{(j)})$ that were used in the calculation of $z^{(k)}$, and the pair that is discarded need not be the one for which the value of j is least. The vector $d^{(j)}$ that is replaced is the ith column of $D^{(k)}$, where i is the integer, $1 \leqslant i \leqslant n$, that gives the largest value of $|z_i^{(k)}\{D^{(k)T}f(x^{(k)})\}_i|$.

Peckham (1970) also includes a linear search process to ensure that the values of $F(x^{(k)})$ ($k = 1, 2, \ldots$) do not diverge, and he finds that his method is more efficient than Powell's (1965) in terms of function evaluations, but it is a little less efficient in terms of the number of routine computing operations per iteration.

Besides the generalized secant method, there is another very useful idea for estimating first derivatives without applying formula (3.3.1), that is due to Broyden (1965) and to Barnes (1965). To describe it we suppose that we have an estimate of the matrix $J^{(k)}$, and that using this estimate $x^{(k+1)}$ has been calculated, possibly by following equation (3.2.6). Then we obtain the estimate $J^{(k+1)}$ (It is now convenient to use the notation $J^{(k)}$ for the *estimate* of the first derivative matrix) from the matrix $J^{(k)}$, and from the vectors $s^{(k)}$ and $d^{(k)}$, defined by equation (3.3.4). If the functions $f_t(x)$ ($t = 1, 2, \ldots, m$) are all linear, and the true first derivative matrix is $\bar{J}$, then the equation

$$\bar{J}s^{(k)} = d^{(k)} \tag{3.3.8}$$

holds. Therefore the idea is to obtain the equation $J^{(k+1)}s^{(k)} = d^{(k)}$ by letting $J^{(k+1)}$ be the matrix

$$J^{(k+1)} = J^{(k)} + \frac{(d^{(k)} - J^{(k)}s^{(k)})w^{(k)T}}{(w^{(k)}, s^{(k)})}, \tag{3.3.9}$$

for some vector $w^{(k)}$.

In fact equation (3.3.9) provides a method for updating the matrix of the secant method. Specifically if $J^{(k)}$ is the matrix corresponding to equations (3.3.2) and (3.2.3), and if we let $w^{(k)}$ be a vector that is orthogonal to $s^{(k-n+1)}$, $s^{(k-n+2)}, \ldots, s^{(k-1)}$, then it follows that the matrix (3.3.9) is the first derivative approximation for the $(k+1)$th iteration of the generalized secant method (Barnes, 1965). Further, equation (3.3.9) strengthens our apprehension about the possibility of numerical difficulties when using the secant method, including perhaps some linear searches to prevent divergence, because if the method calculates a sequence of points $x^{(k)}$ $(k = 1, 2, \ldots)$ that converge to the position of the least value of $F(x)$, and if the condition $F(x^{(k+1)}) \leqslant F(x^{(k)})$ $(k = 1, 2, \ldots)$ holds, then usually the sequence of points $x^{(k)}$ includes some trends. Therefore the direction $w^{(k)}$, that is orthogonal to $s^{(k-n+1)}$, $s^{(k-n+2)}, \ldots, s^{(k-1)}$, is liable to be orthogonal to $s^{(k)}$ also, so it is not satisfactory to use equation (3.3.9).

Broyden (1965) therefore considers other choices for the vector $w^{(k)}$ in equation (3.3.9), and he recommends an algorithm that sets $w^{(k)} = s^{(k)}$ for all $k \geqslant 1$. However, this algorithm does not have the property of exactly minimizing $F(x)$ in a finite number of iterations if the functions $f_t(x)$ $(t = 1, 2, \ldots, m)$ are linear. Instead it has a different nice property when all the functions are linear, namely that equations (3.3.8) and (3.3.9) in the case $w^{(k)} = s^{(k)}$ imply the identity

$$J^{(k+1)} - \bar{J} = \{J^{(k)} - \bar{J}\}\left\{I - \frac{s^{(k)}s^{(k)T}}{\|s^{(k)}\|^2}\right\}, \tag{3.3.10}$$

so on each iteration the error of the first derivative approximation is reduced by multiplication by a symmetric projection matrix. It is therefore not surprising that Broyden's algorithm works well in practice, and numerical examples indicate that usually it is faster and more reliable than the algorithms that are based on the generalized secant method.

We mentioned that instead of solving equations to compute the vector $z^{(k)}$ of equation (3.3.7), it is more efficient, in terms of the number of computer operations, to store the matrix $\{D^{(k)T}D^{(k)}\}^{-1}$, and to calculate $\{D^{(k+1)T}D^{(k+1)}\}^{-1}$ by a recurrence formula (Rosen, 1960). Similar savings in the amount of computation can be made in Broyden's (1965) algorithm. For example, if $m = n$, and if $J^{(k)}$ is non-singular, then equation (3.2.12) reduces to $p^{(k)} = -[J^{(k)}]^{-1} f(x^{(k)})$, so it may be worthwhile to store the matrix $[J^{(k)}]^{-1}$ and to calculate $[J^{(k+1)}]^{-1}$ by applying Householder's formula to equation (3.3.9) (Broyden, 1965). One possible difficulty is singularity of the matrix $J^{(k+1)}$, but otherwise this calculation can be carried out in a numerically stable way (Powell, 1969b). In the case that $m > n$ one may update the matrix $[J^{(k)T}J^{(k)}]^{-1}$, or a factorization of $[J^{(k)T}J^{(k)}]$ (Bennett, 1965; Bartels, Golub and Saunders, 1970), or one may find an updating formula for the matrix $[J^{(k)T}J^{(k)}]^{-1}J^{(k)T}$. Thus only of order mn computer operations per iteration are needed by many of the algorithms that estimate first derivatives. However, in section 3.2 we noted that if first derivatives are calculated directly, then of order mn^2 operations per iteration are required by the algorithms that minimize a sum of squares.

So far in this section we have discussed only the problem of estimating derivatives, because, given an approximation to the first derivative matrix, most of the remarks of section 3.2 hold. For instance, to prevent divergence one frequently needs to include a parameter that can be adjusted to give a reduction in the objective function on each iteration, $F(x^{(k+1)}) < F(x^{(k)})$ $(k = 1, 2, \ldots)$. Broyden (1965) uses a parameter $\alpha^{(k)}$ in the way shown by equations (3.2.11), (3.2.12) and (3.2.13), but we have noted that there are advantages in using the Marquardt formula (3.2.14) to define $p^{(k)}$. Therefore Powell (1970b) has developed a method for solving systems of non-linear equations, that combines formula (3.3.9) with the idea of the Marquardt algorithm.

There is a difficulty in trying to follow Marquardt's idea, which is that if the value of $\lambda^{(k)}$ is altered in equation (3.2.14), then a new set of linear equations has to be solved to define $p^{(k)}$, so a direct application of the idea necessitates of order n^3 operations per iteration. We have noted already that this amount of computing would be excessive in an algorithm that estimates derivatives, so Powell follows only the essence of Marquardt's idea, which is that if $\lambda^{(k)} = 0$ then equation (3.2.12) holds, and if $\lambda^{(k)}$ becomes large then the length of $p^{(k)}$ decreases, and its direction tends to be along the steepest-descent vector of $F(x)$ at $x^{(k)}$.

Specifically, instead of letting $p^{(k)}$ be the vector (3.2.14) for some $\lambda^{(k)}$, Powell lets $p^{(k)}$ be a displacement from $x^{(k)}$ to a point on a "dog-leg". The locus of the end of $p^{(k)}$ is initially down the steepest-descent vector at $x^{(k)}$, and then at a certain point on this vector, the locus switches to the straight line to the point (3.2.12). The switching point, that is the point at which the dog-leg bends, is the point

$$q^{(k)} = -\frac{\|J^{(k)T}f(x^{(k)})\|^2}{\|J^{(k)}J^{(k)T}f(x^{(k)})\|^2}\, J^{(k)T}f(x^{(k)}), \tag{3.3.11}$$

because the least value of the function (3.2.4), subject to the condition that x is on the steepest-descent vector from $x^{(k)}$, occurs when $x = x^{(k)}+q^{(k)}$. Therefore in Powell's (1970b) algorithm the matrices $J^{(k)}$ and $[J^{(k)}]^{-1}$ are both stored.

Jones (1970) suggests another way of keeping the essence of Marquardt's idea, without having to solve a set of linear equations to adjust $p^{(k)}$. Like Powell, Jones forces $p^{(k)}$ to be in the two-dimensional space containing the point (3.2.12) and the gradient of $F(x)$ at $x^{(k)}$, but, instead of using a dog-leg, Jones uses a curve whose slope is continuous to define the possible values of $p^{(k)}$. Of course the initial direction of this curve is along the steepest-descent direction at $x^{(k)}$, and the curve bends so that it reaches the point (3.2.12).

In these two sections on least squares algorithms it has been stated that the methods that are the most reliable at the present time include a bias towards the steepest-descent direction. This is unsatisfactory because it implies that the user of one of these algorithms has to ensure that the components of x are scaled so that they do not differ very much in magnitude. Instead we would like to have algorithms that are robust, and that scale the variables automatically—a worthwhile research subject. Another worthwhile subject for research is to investigate the improvements to current algorithms that can be obtained by storing matrices in factored form (Bartels *et al.*, 1970).

3.4 Transforming constrained into unconstrained problems

In the final three sections of this paper we consider the problem of calculating the least value of a given function $F(x)$, subject to constraints on the variables, $a_i(x) \geqslant 0$, $i = 1, 2, \ldots, \alpha$ and $c_i(x) = 0$, $i = 1, 2, \ldots, \gamma$ say. There are a number of methods that treat the constraints in a direct way (see Davies and Swann, 1969, for instance), but, because this paper is on problems related to unconstrained optimization, we consider only those methods that solve constrained problems by using an algorithm for unconstrained optimization.

For example, let us suppose that we wish to calculate the least value of $F(x)$ subject to $x_i \geqslant 0$ $(i = 1, 2, \ldots, n)$. Then if we define $\bar{F}(w)$ to be the function

$$\bar{F}(w_1, w_2, \ldots, w_n) = F(w_1^2, w_2^2, \ldots, w_n^2), \tag{3.4.1}$$

and if we calculate the least value of $\bar{F}(w)$ by applying an algorithm for unconstrained optimization, then from equation (3.4.1) it is clear that we will have calculated the least value of $F(w_1^2, w_2^2, \ldots, w_n^2)$, which is the least value of $F(x)$ subject to the given constraints, $x_i \geqslant 0$ $(i = 1, 2, \ldots, n)$. Further, the required value of x_i $(i = 1, 2, \ldots, n)$ is the square of the final value of w_i. This idea was proposed by Dickinson (1964) and by Box (1966).

Further, if in addition to the non-negativity constraints $x_i \geqslant 0$ $(i = 1, 2, \ldots, n)$, we have the upper bound $x_1 \leqslant 1$, then we can take account of this extra constraint by defining $\bar{F}(w)$ to be the function

$$\bar{F}(w_1, w_2, \ldots, w_n) = F(\sin^2 w_1, w_2^2, w_3^2, \ldots, w_n^2) \tag{3.4.2}$$

instead of the function (3.4.1). Because we let $x_1 = \sin^2 w_1$, the constraint $0 \leqslant x_1 \leqslant 1$ is satisfied automatically, so now we calculate the required least value of $F(x)$ by applying an algorithm for unconstrained optimization to the function (3.4.2).

In both these examples we have found transformations $x_i(w_i)$ $(i = 1, 2, \ldots, n)$, such that if w_i is any real number, then $x_i(w_i)$ satisfies the constraints on x_i. Another important property of these transformations is that for every feasible value of x_i, there exists a value of w_i such that $x_i = x_i(w_i)$. We note that the function that is minimized is $\bar{F}(w) = F(x(w))$.

It is straightforward to generalize this idea in a useful way. Specifically if the constraints on x are $x \in R$, R being a region of Euclidean n-space, then we look for a function $x(w)$ that satisfies two conditions: (i) for all real vectors w, $x(w) \in R$, and (ii) for every $x \in R$ there exists at least one real vector w such that $x = x(w)$. For example, if $n = 2$, and there is one constraint on the variables, namely $x_1^2 + x_2^2 \leqslant 1$, then it is satisfactory to make the transformation

$$\left.\begin{aligned} x_1 &= \sin w_1 \sin^2 w_2 \\ x_2 &= \cos w_1 \sin^2 w_2 \end{aligned}\right\}. \tag{3.4.3}$$

There is no need for the number of components of w to equal n, and usually the number of components of w equals n minus the number of equality constraints. For example, if $n = 3$, and the constraints are $-1 \leqslant x_1 \leqslant 1$ and $x_2 = 5x_3$, then we let w have two components, and we use the transformation

$$\left.\begin{aligned} x_1 &= \sin w_1 \\ x_2 &= w_2 \\ x_3 &= 0{\cdot}2 w_2 \end{aligned}\right\}. \tag{3.4.4}$$

However, a strong disadvantage of the method of this section is that frequently the constraints are so complicated or so numerous that it is not possible to find an adequate transformation of the variables.

3.5 Penalty function methods for constraints

One of the first penalty function methods was proposed by Carroll (1961). He considered the problem: calculate the least value of $F(x)$ subject to the constraints $a_i(x) \geqslant 0$ $(i = 1, 2, \ldots, \alpha)$, and to solve it he defined the function

$$\Phi(x, \rho) = F(x) + \rho \sum_{i=1}^{\alpha} \frac{1}{a_i(x)}, \tag{3.5.1}$$

where ρ is a small positive parameter. An important feature of the function $\Phi(x, \rho)$ is that it is infinite at every point x that is on a constraint boundary. Therefore if an algorithm for *unconstrained* optimization is applied to the function $\Phi(x, \rho)$, starting at a point $x^{(1)}$ that satisfies the strict inequalities $a_i(x^{(1)}) > 0$ $(i = 1, 2, \ldots, \alpha)$, then, for the calculated sequence of points $x^{(k)}$ $(k = 1, 2, \ldots)$, it should be possible to maintain the inequalities

$a_i(x^{(k)}) > 0$. Certainly if the function (3.5.1) is changed to the function

$$\Phi(x, \rho) = F(x) + \rho \sum_{i=1}^{\alpha} \frac{1}{\max[0, a_i(x)]}, \tag{3.5.2}$$

and if the unconstrained algorithm ensures that no iteration increases the objective function, then feasibility of the calculated points $x^{(k)}$ $(k = 1, 2, \ldots)$ is assured. We let $x(\rho)$ be the final vector of variables calculated by the algorithm for unconstrained minimization.

Because the constraint terms of the functions (3.5.1) and (3.5.2) can be made smaller by decreasing ρ, Carroll (1961) suggests that $x(\rho)$ be calculated for a sequence of positive values of ρ that tends to zero. Then usually $x(\rho)$ converges to the solution of the original constrained problem (Fiacco and McCormick, 1968).

Note that in this method all of the calculated vectors $x^{(k)}$ satisfy the inequalities $a_i(x^{(k)}) > 0$ $(i = 1, 2, \ldots, \alpha)$. Therefore it is not suitable for equality constraints. To solve the problem: minimize $F(x)$ subject to $a_i(x) \geqslant 0$ $(i = 1, 2, \ldots, \alpha)$ and $c_i(x) = 0$ $(i = 1, 2, \ldots, \gamma)$, Fiacco and McCormick (1966) propose the minimization of the function

$$\Phi(x, \rho) = F(x) + \rho \sum_{i=1}^{\alpha} \frac{1}{\max[0, a_i(x)]} + \rho^{-\frac{1}{2}} \sum_{i=1}^{\gamma} [c_i(x)]^2, \tag{3.5.3}$$

where again one uses a sequence of values of ρ that tends to zero. A property of formula (3.5.3) is that as ρ tends to zero, the penalty for failing to satisfy an equality constraint, $c_i(x) = 0$, becomes larger and larger, so in this case also $x(\rho)$ usually converges to the required limit.

Now, guided by expressions (3.5.1), (3.5.2) and (3.5.3), we give a general definition of a penalty function method. To solve the problem of minimizing $F(x)$ subject to $x \in R$, we find a function $\psi(x, \rho)$, where ρ is a positive parameter, that satisfies certain conditions. The most important ones are that if x is not in R and ρ tends to zero, then $\psi(x, \rho)$ tends to infinity, and, for most values of x in R, if ρ tends to zero, then $\psi(x, \rho)$ tends to zero. This last statement is vague because the penalty function of expression (3.5.2) tends to infinity when x is on the boundary of R. Then in a penalty function method the vector of variables, $x(\rho)$ say, that minimizes the function

$$\Phi(x, \rho) = F(x) + \psi(x, \rho) \tag{3.5.4}$$

is calculated by an algorithm for unconstrained optimization for a sequence of values of ρ that tends to zero, because usually $x(\rho)$ converges to the solution of the constrained problem.

It is possible to write down conditions on $F(x)$ and $\psi(x, \rho)$ that guarantee the success of the penalty function method (assuming that a perfect algorithm for unconstrained minimization is available!), but they are too sophisticated for consideration here. However, it is worthwhile to note that these conditions do not depend on convexity or on differentiability, so penalty function methods are applicable to most constrained problems. Some theorems on convergence when ρ tends to zero are proved by Fiacco and McCormick (1968) and by Lootsma (1970).

Besides the function of equation (3.5.1), two other functions are used frequently for solving the problem: minimize $F(x)$ subject to the constraints $a_i(x) \geqslant 0$ $(i = 1, 2, \ldots, \alpha)$. They are the functions

$$\psi(x, \rho) = \begin{cases} -\rho \sum_{i=1}^{\alpha} \log a_i(x), & x \in R \\ \infty, & x \notin R \end{cases} \tag{3.5.5}$$

and

$$\psi(x, \rho) = \rho^{-\frac{1}{2}} \sum_{i=1}^{\alpha} [\min \{0, a_i(x)\}]^2. \tag{3.5.6}$$

There is no need in theory to square the terms under the summation sign of equation (3.5.6), but in practice this is worthwhile, to avoid discontinuities in the gradient of $\psi(x, \rho)$ on the boundary of R.

One necessary condition for the success of penalty function methods that should be kept in mind is that it must be possible to minimize the function (3.5.4). Therefore if $F(x)$ can tend to minus infinity for a sequence of values of x that violates the constraints, then $\psi(x, \rho)$ must be so large that it dominates the very negative values of $F(x, \rho)$. For example, although the problem of minimizing x^3 subject to $x \geqslant 1$ has a solution, it is not adequate to let

$$\Phi(x, \rho) = x^3 + \rho^{-\frac{1}{2}}[\min \{0, x-1\}]^2, \tag{3.5.7}$$

because $\Phi(x, \rho)$ tends to minus infinity as x tends to minus infinity for all $\rho > 0$. Even when $F(x)$ is bounded below the introduction of penalty functions can cause difficulties. For example, if to minimize the function $-1/(1+x^2)$ subject to $x \geqslant 1$, we let

$$\Phi(x, \rho) = \frac{-1}{1+x^2} - \rho \log (x-1), \tag{3.5.8}$$

then $\Phi(x, \rho)$ tends to minus infinity as x tends to infinity for all $\rho > 0$. However, if $F(x)$ is bounded below, and if $\psi(x, \rho) \geqslant 0$, then usually the difficulties shown in this paragraph do not occur.

In some books it is stated that the success of a penalty function method depends on convexity conditions, but this is not true. However, if certain convexity conditions hold then one can prove some useful duality theorems. We now give an example of one of these theorems (Fiacco and McCormick, 1968).

If the problem is to minimize $F(x)$ subject to $a_i(x) \geqslant 0$ $(i = 1, 2, \ldots, \alpha)$, if $F(x)$ is a convex function, if all the functions $a_i(x)$ are concave, and if

$x(\rho)$ minimizes the function (3.5.2), then not only does the inequality $F(x(\rho)) \geqslant F(x^*)$ hold, where x^* is the vector of variables that solves the problem, but also the inequality

$$F(x(\rho))-\rho \sum_{i=1}^{\alpha} \frac{1}{a_i(x(\rho))} < F(x^*) \tag{3.5.9}$$

is satisfied. Thus each calculation of $x(\rho)$ provides upper and lower bounds on $F(x^*)$, which are useful because as ρ tends to zero, the expression $\rho\ \Sigma\ 1/a_i(x(\rho))$ also tends to zero.

We now suppose that a suitable penalty function $\psi(x, \rho)$ has been found, and we consider some choices that have to be made in order to write a computer program.

An unconstrained minimization algorithm is required to calculate the least value of $\Phi(x, \rho)$. Almost certainly it will be one of the iterative methods described in this volume, so it will require an initial vector of variables, $x^{(1)}$ say. Note that if $\Phi(x, \rho)$ is defined by equation (3.5.2), (3.5.3) or (3.5.5), then $x^{(1)}$ has to satisfy the inequality constraints. Usually one can spot a suitable vector $x^{(1)}$, but if a feasible vector has to be calculated, then one may apply an algorithm for unconstrained optimization to the function

$$\sum_{i=1}^{\alpha} [\min\{0, a_i(x)-\varepsilon\}]^2, \tag{3.5.10}$$

where ε is a small positive constant to make $a_i(x^{(1)})$ strictly positive for $i = 1, 2, \ldots, \alpha$.

Next an initial value has to be assigned to ρ. We would like the term $\psi(x, \rho)$ to be comparable to the term $F(x)$, and, given differentiable functions, we know that the gradient of $\Phi(x, \rho)$ at $x(\rho)$ should be zero. Therefore Fiacco and McCormick (1964) suggest that the initial value of ρ be calculated to minimize the gradient of $\Phi(x, \rho)$ at $x = x^{(1)}$. But it may happen that this value of ρ is not positive, in which case we let $\rho = 0$, and we apply an algorithm for unconstrained minimization to $F(/x)$, starting at $x^{(1)}$. This may lead to the solution of the problem, but usually it will lead to a point x at which a positive initial value of ρ can be calculated by minimizing the gradient of $\Phi(x, \rho)$. Although there are defects in this scheme, it is recommended in many papers on penalty functions, because it can be programmed so that the initial value of ρ is assigned automatically.

Another decision is to fix the later values of ρ, so that we have a sequence of positive values that converges to zero. Fiacco and McCormick (1968) suggests that ρ be divided by a constant, β say, at each stage, and they claim that the actual choice of β is not very critical. Kowalik and Osborne (1968) state that in practice they have found it adequate to let $\beta = 10$.

Of course, the unconstrained minimizations of $\Phi(x, \rho)$ have to be terminated, so another consideration is the accuracy to which $x(\rho)$ is calculated.

There seems to be no advantage in calculating $x(\rho)$ accurately if it is used only as a starting vector for the next unconstrained minimization. However, many algorithms for unconstrained minimization do have a superlinear rate of convergence, so obtaining high accuracy at each stage need not be expensive. I have not read a published paper that includes a good discussion of the advantages and disadvantages of minimizing $\Phi(x, \rho)$ accurately for each ρ.

However, in one case it is important to obtain high accuracy in the calculation of $x(\rho)$, which is when an "extrapolation technique" is used to predict x^*. Here is an example.

To solve the problem of calculating the least distance from the point (0, 0) to the curve $y = e^x$, we minimize the function $F(x_1, x_2) = x_1^2+x_2^2$, subject to the constraint $x_2-e^{x_1} \geqslant 0$, by a penalty function method. Specifically we calculate the least value of the function

$$\Phi(x, \rho) = x_1^2+x_2^2+\rho/\max\,[0, x_2-e^{x_1}] \qquad (3.5.11)$$

for some values of ρ, and we obtain the results given in Table 3.I. We require the limiting values of $x_1(\rho)$, $x_2(\rho)$ and $F(x(\rho))$ as ρ tends to zero.

Because the numbers in Table 3.I are correct to at least five decimals, there is no need to solve a further unconstrained minimization problem

TABLE 3.I. *Minimization of the function* (3.5.11)

ρ	$x_1(\rho)$	$x_2(\rho)$	$\Phi(x(\rho))$
1·0	−0·629965	1·18279	3·33388
0·1	−0·508224	0·844839	1·38310
0·01	−0·455257	0·717749	0·842235
0·001	−0·435853	0·673951	0·680891
0·0001	−0·429366	0·659628	0·630950

to improve on the accuracy of $x(0{\cdot}0001)$. Instead we extrapolate using the data given in the table in the following way. It can be shown that the dependence of $x_i(\rho)$ on ρ has the form

$$x_i(\rho) = x_i^*+u_i\rho^{\frac{1}{2}}+v_i\rho+0(\rho^{\frac{3}{2}}) \qquad (3.5.12)$$

(Fiacco and McCormick, 1966), and therefore we define

$$y_i(\rho) = \frac{x_i(\rho)-\sqrt{0{\cdot}1}\,x_i(10\rho)}{1-\sqrt{0{\cdot}1}} \qquad (3.5.13)$$

and

$$z_i(\rho) = \frac{y_i(\rho)-0{\cdot}1y_i(10\rho)}{0{\cdot}9}, \qquad (3.5.14)$$

because it follows that the difference between x^* and $z(\rho)$ is of order $\rho^{3/2}$. Therefore we calculate the vector $z(0{\cdot}0001)$, for we expect it to be better than $x(0{\cdot}0001)$ as an estimate of x^*. This calculation is shown in Table 3.II. In fact x^* is the vector $(-0{\cdot}426303, 0{\cdot}652919)$, so the extrapolation has been successful, for it has reduced the error by about the factor 1000.

TABLE 3.II. *Extrapolation to x^**

ρ	$x_1(\rho)$	$y_1(\rho)$	$z_1(\rho)$
1·0	−0·629965		
0·1	−0·508224	−0·451922	
0·01	−0·455257	−0·430761	−0·428410
0·001	−0·435853	−0·426879	−0·426448
0·0001	−0·429366	−0·426366	−0·426309

ρ	$x_2(\rho)$	$y_2(\rho)$	$z_2(\rho)$
1·0	1·18279		
0·1	0·844839	0·688545	
0·01	0·717749	0·658973	0·655687
0·001	0·673951	0·653696	0·653110
0·0001	0·659628	0·653004	0·652927

Given that the functions $F(x)$ and $a_i(x)$ of equation (3.5.2) are differentiable. expression (3.5.12) is usually correct, and this expression is also relevant to equation (3.5.3), because of the factor $\rho^{-\frac{1}{2}}$ multiplying $\Sigma\,[c_i(x)]^2$. For further details, including an analysis of the trajectory of $x(\rho)$ when the penalty functions (3.5.5) and (3.5.6) are used, see the book by Fiacco and McCormick (1968).

There is no doubt that the extrapolation procedure, shown in Table 3.II is well worthwhile for improving an estimate of x^*, but also it has other uses. For instance, because we calculate the point $x(0{\cdot}0001)$ of Table 3.I by setting $\rho = 0{\cdot}0001$ and by applying an unconstrained minimization algorithm to the function (3.5.11), we require an initial estimate of the vector $x(0{\cdot}0001)$. We could let this estimate be the vector $x(0{\cdot}001)$, but instead it is better to extrapolate from the first four lines of Table 3.I. Specifically, guided by equation (3.5.12), we note that the vector

$$w(\rho) = \begin{pmatrix} -0{\cdot}426448 - 0{\cdot}301726\sqrt{\rho} + 0{\cdot}13638\rho \\ 0{\cdot}653110 + 0{\cdot}664910\sqrt{\rho} - 0{\cdot}18534\rho \end{pmatrix} \tag{3.5.15}$$

is equal to $x(\rho)$ at $\rho = 0{\cdot}1$, $0{\cdot}01$ and $0{\cdot}001$. Therefore a good initial estimate of $x(0{\cdot}0001)$ is the vector $w(0{\cdot}0001)$, whose components are $(-0{\cdot}429452, 0{\cdot}659741)$, which demonstrates that extrapolation can also be used to help the unconstrained minimizations.

Many of the algorithms for unconstrained minimization require the user to provide first derivatives, and an initial estimate of the second derivative matrix, and then the second derivative estimates are refined automatically. If one of these algorithms is applied, then, as well as using the method of the last paragraph to estimate $x(\rho)$, we should try to provide an equally good estimate of the required second derivative matrix. Therefore Fletcher and McCann (1969) have analysed the dependence on ρ of the second derivative matrix of $\Phi(x, \rho)$ at the point $x = x(\rho)$ for the penalty function (3.5.2). They find that it tends to be of the form $A+\rho^{-\frac{1}{2}}B$, where A and B are independent of ρ, and where the rank of B is equal to the number of constraints $a_i(x) \geqslant 0$ $(i = 1, 2, \ldots, \alpha)$ that become equalities at $x = x^*$. Their analysis leads to a good method for estimating the matrix that is required by Davidon's (1959) algorithm for unconstrained minimization.

All these remarks on extrapolation, and on analysing the trajectory of $x(\rho)$, lead one to the conclusion that it may be wrong to regard penalty function methods as a sequence of discrete unconstrained minimizations but instead one should think of ρ as a quantity that varies continuously (Murray, 1969). This approach is very promising, but it takes us beyond the scope of this chapter, namely problems related to unconstrained optimization.

3.6 Lagrange parameter methods

A feature of penalty function methods is that as the quantity ρ of equation (3.5.4) tends to zero, the function $\psi(x, \rho)$ tends to have an infinite discontinuity on the constraint boundaries, so it is usually awkward to calculate the least value of the function $\Phi(x, \rho)$ by a numerical algorithm for unconstrained minimization. Therefore it is worthwhile to look for other methods for treating constraints on the variables. Many other methods have been proposed, and a few of them depend mainly on the availability of a good algorithm for unconstrained minimization. These few methods are the subject of this section.

At the present time these methods do not treat inequality constraints so in this section we suppose that the problem to be solved is: calculate the least value of $F(x)$ subject to $c_i(x) = 0$ $(i = 1, 2, \ldots, \gamma)$. If inequalities are present, then somehow one has to isolate the inequality constraints that become equalities at the required point x^*, in order that one can regard just these inequality constraints as equalities throughout the calculation.

One method for solving the constrained problem is the method of Lagrange multipliers. Here we let $\Lambda(x, l)$ be the function

$$\Lambda(x, l) = F(x) + \sum_{i=1}^{\gamma} l_i c_i(x), \tag{3.6.1}$$

where the l_i ($i = 1, 2, \ldots, \gamma$) are the "Lagrange parameters". This method depends on the well-known theorem that, if the functions $F(x)$ and $c_i(x)$ are differentiable, then there exist values l_i^* ($i = 1, 2, \ldots, \gamma$), such that (x^*, l^*) is a stationary point of the function $\Lambda(x, l)$. Specifically we look for vectors (x, l) that satisfy the $(n+\gamma)$ equations

$$c_i(x) = 0 \quad (i = 1, 2, \ldots, \gamma) \tag{3.6.2}$$

and

$$\frac{\partial}{\partial x_j} F(x) + \sum_{i=1}^{\gamma} l_i \frac{\partial c_i(x)}{\partial x_j} = 0 \quad (j = 1, 2, \ldots, n). \tag{3.6.3}$$

This method is frequently used in algebraic work, and it has been developed into a numerical algorithm by Broyden and Hart (1970). A practical difficulty is that equations (3.6.2) and (3.6.3) have solutions not only at constrained minima of $F(x)$, but also at constrained saddle points and at constrained maxima.

To avoid constrained maxima, it is good to use a method that tends to make $F(x)$ small, so we wish to minimize a function like expression (3.6.1). However, this function depends linearly on l_i ($i = 1, 2, \ldots, \gamma$), so we may not regard the Lagrange parameters as independent variables. Instead we let l be an estimate of l^*, and without varying l, we try to calculate the vector x that minimizes expression (3.6.1), say it is $x(l)$. Then, guided by $x(l)$, we try to improve our estimate of l^*, and, for a new value of l, we calculate the new least value of $\Lambda(x, l)$. If we continue this process successfully, we obtain a sequence of vectors $x(l)$ that converges to x^*, by solving a sequence of unconstrained minimization problems. Therefore this method is similar to the penalty function methods, however now it is usually possible to find an algorithm that does not introduce tendencies to infinite discontinuities in the functions that are minimized.

The main defect of the method of the last paragraph is that when $l = l^*$, x^* need not minimize the function $\Lambda(x, l)$. For example if we try to calculate the least value of x^3 subject to $x = 1$, we find the function

$$\Lambda(x, l^*) = x^3 - 3x, \tag{3.6.4}$$

which tends to infinity as x tends to minus infinity (compare equation (3.5.7)). Therefore, as in section 3.5, we have to be careful if $F(x)$ can tend to minus infinity for vectors x that violate the constraints. Also, as in equation (3.5.8), it can happen that the Lagrange parameter method for converting a constrained problem into an unconstrained one can introduce difficulties even when $F(x)$ is bounded below. For example, if $F(x) = x_1+x_2+\frac{1}{2}(x_1^2+x_2^2)$, and if there is only one constraint, namely $(x_1+x_1^2+x_1^3)/(1+x_1^4) = 0$, then we have the function

$$\Lambda(x, l^*) = x_1 + x_2 + \tfrac{1}{2}(x_1^2+x_2^2) - \frac{x_1+x_1^2+x_1^3}{1+x_1^4}. \tag{3.6.5}$$

In this case the required point $x^* = (0, -1)$ is not a minimum of $\Lambda(x, l^*)$ but it is a saddle point.

A useful way of avoiding the difficulty shown by the last example is to modify the function (3.6.1). For example we define the function

$$\Theta(x, l, h) = F(x) + \sum_{i=1}^{\gamma} l_i c_i(x) + \sum_{i=1}^{\gamma} h_i[c_i(x)]^2. \tag{3.6.6}$$

If we regard the components of x, l and h as independent variables, then by differentiation it follows that (x, l, h) is a stationary point of $\Theta(x, l, h)$ if and only if equations (3.6.2) and (3.6.3) hold. Therefore if, for $i = 1, 2, \ldots, \gamma$, h_i is given any constant value, then x^* is a stationary point of the function $\Theta(x, l^*, h)$. This remark is useful because at $x = x^*$, the second derivative matrix with respect to x of the last term of equation (3.6.6) is positive semidefinite, and therefore by assigning sufficiently large values to h_i $(i = 1, 2, \ldots, \gamma)$, we can usually force x^* to be a minimum of the function $\Theta(x, l^*, h)$. For example, although x^* is a saddle point of expression (3.6.5), it is a minimum of the function

$$x_1 + x_2 + \tfrac{1}{2}(x_1^2+x_2^2) - \frac{x_1+x_1^2+x_1^3}{1+x_1^4} + \left\{\frac{x_1+x_1^2+x_1^3}{1+x_1^4}\right\}^2. \tag{3.6.7}$$

Therefore it is worthwhile to modify the method outlined between equations (3.6.3) and (3.6.4) so that, if as l is adjusted the values of the constraints $c_i(x(l))$ $(i = 1, 2, \ldots, \gamma)$ fail to tend to zero, then multiples of $[c_i(x)]^2$ are added into the function that is treated by the algorithm for unconstrained minimization.

This feature is present in Powell's (1969a) algorithm for minimization subject to equality constraints, and his method is as follows. It is iterative, and for the kth iteration $(k = 1, 2, \ldots)$ there are two vectors of parameters, namely $l^{(k)}$ and $h^{(k)}$. The main operation of this iteration is to minimize the function $\Theta(x, l^{(k)}, h^{(k)})$, defined by equation (3.6.6), and we let $x^{(k)}$ be the resultant vector of variables. Then $l^{(k+1)}$ and $h^{(k+1)}$ are calculated for the next iteration. Usually they are defined by the formula

$$\left.\begin{aligned} l_i^{(k+1)} &= l_i^{(k)} + 2h_i^{(k)} c_i(x^{(k)}) \\ h_i^{(k+1)} &= h_i^{(k)} \end{aligned}\right\} (i = 1, 2, \ldots, \gamma), \tag{3.6.8}$$

but if the constraint functions $c_i(x)$ are not converging to zero well, then $h^{(k+1)}$ is made larger than $h^{(k)}$. Specifically the inequality

$$\max_{1 \leqslant i \leqslant \gamma} |c_i(x^{(k)})| \leqslant \tfrac{1}{4} \max_{1 \leqslant i \leqslant \gamma} |c_i(x^{(k-1)})| = \eta^{(k-1)}, \tag{3.6.9}$$

say, is tested, and if it holds then formula (3.6.8) is used. Otherwise the new parameter vectors are defined by the equation

$$\left.\begin{array}{l} l_i^{(k+1)} = l_i^{(k)} \\ h_i^{(k+1)} = h_i^{(k)},\ |c_i(x^{(k)})| < \eta^{(k-1)} \\ h_i^{(k+1)} = 10h_i^{(k)},\ |c_i(x^{(k)})| \geqslant \eta^{(k-1)} \end{array}\right\}. \tag{3.6.10}$$

However, when $k = 1$ the inequality (3.6.9) is inapplicable, and therefore equation (3.6.8) is always used to define $l^{(2)}$ and $h^{(2)}$. The user of the method has to assign the values of $l^{(1)}$ and $h^{(1)}$, and every component of $h^{(1)}$ must be positive. For example the values $l_i^{(1)} = 0$, $h_i^{(1)} = 1$ ($i = 1, 2, \ldots, \gamma$) may be suitable.

Powell studies the convergence of his method, and finds that in addition to continuity and boundedness conditions, it is necessary for the functions $c_i(x)$ ($i = 1, 2, \ldots, \gamma$) to satisfy an independence condition at x^*. For example it is adequate if the $nx\gamma$ matrix $N(x)$, whose (i, j)th component is

$$N_{ij}(x) = \partial c_j(x)/\partial x_i, \tag{3.6.11}$$

has rank γ at $x = x^*$. Then it is proved that the method converges at a linear rate, the geometric factor being at most $\frac{1}{4}$. However, for any $\delta > 0$, a geometric factor of at most δ can be obtained by replacing $\frac{1}{4}$ by δ in inequality (3.6.9). Unfortunately if δ is very small, then the components of $h^{(k)}$ tend to be very large, in which case the unconstrained minimizations become more difficult, so we revert to the disadvantages of penalty function methods.

Hestenes (1969) also shows that the function (3.6.6) is useful for minimizing $F(x)$ subject to $c_i(x) = 0$ ($i = 1, 2, \ldots, \gamma$).

Expression (3.6.6) is just one of many ways of extending the Lagrangian function (3.6.1). In general we can apply an algorithm for unconstrained minimization to the function

$$\Theta(x, h) = F(x) + \psi(c_1, c_2, \ldots, c_\gamma, h), \tag{3.6.12}$$

where h is a vector of parameters (which may include the Lagrange parameters l_i, $i = 1, 2, \ldots, \gamma$), and where ψ is any convenient function that depends on the parameters and on the values of the constraint functions. Note that x is not an explicit argument of ψ but that ψ does depend on x through the functions $c_i(x)$ ($i = 1, 2, \ldots, \gamma$). Let $x(h)$ be the vector of variables that minimizes expression (3.6.12). Then the following statement is true (Powell, 1969a).

$x(h)$ solves the constrained optimization problem: minimize $F(x)$ subject to the equality constraints $c_i(x) = c_i(h)$ ($i = 1, 2, \ldots, \gamma$), where $c_i(h) = c_i(x(h))$. To prove this statement, we suppose that it is false, and that $y \neq x(h)$ solves this problem. Then $F(y) < F(x(h))$, and it follows that $\Theta(y, h) < \Theta(x(h), h)$, which is a contradiction.

Although this last statement is almost obvious, it is very important, because it implies that whenever we calculate the least value of a function of the form (3.6.12), then we do solve a constrained problem. Further, to solve the original problem, we now know that we just have to adjust the parameter vector h so that the numbers $c_i(h)$ $(i = 1, 2, \ldots, \gamma)$ become zero, or tolerably small. This remark could lead to many different numerical methods for constrained optimization problems.

The methods that have been discussed so far in sections 3.5 and 3.6 reduce a constrained minimization problem to a *sequence* of unconstrained calculations, but we would prefer methods that require an algorithm for unconstrained minimization to be applied only once. Recently Fletcher (1970a), Haarhoff and Buys (1970) and Miele *et al.* (1971) have developed such methods, based on Lagrange parameters, and Fletcher's method has solved many test problems very well (Fletcher and Lill, 1970). Therefore it is an important advance in the subject of optimization. This method will now be described.

We have noted already that if we are going to use an algorithm for unconstrained minimization, then the parameters l_i $(i = 1, 2, \ldots, \gamma)$ of equation (3.6.1) cannot be treated as independent variables, and one can hardly ever predict their optimal values *ab initio*. Therefore Fletcher (1970a) lets l_i $(i = 1, 2, \ldots, \gamma)$ depend on x. At x^* the optimal parameter values satisfy equation (3.6.3), so this equation is used to define $l_i(x)$. It is assumed that for every x the matrix $N(x)$ (see equation (3.6.11)) has rank γ, which is not unduly restrictive because usually $\gamma < n$. Then $l(x)$ is defined to be the least squares solution of the overdetermined system of linear equations

$$\sum_{j=1}^{\gamma} N_{ij}(x) l_j(x) + \frac{\partial}{\partial x_i} F(x) = 0 \quad (i = 1, 2, \ldots, n). \tag{3.6.13}$$

We note that the definition of $l(x)$ depends on the first derivatives of $F(x)$ and $c_i(x)$ $(i = 1, 2, \ldots, \gamma)$.

In place of expression (3.6.1) we use the function

$$\Phi_0(x) = F(x) + \sum_{i=1}^{\gamma} l_i(x) c_i(x). \tag{3.6.14}$$

By straightforward differentiation it can be verified that x^* is a stationary point of $\Phi_0(x)$, so there is a possibility that x^* can be obtained by calculating the least value of $\Phi_0(x)$. However, it frequently happens that x^* is a saddle point of $\Phi_0(x)$. For example, if the problem is to minimize $x_1^2 + x_2^2$ subject to $x_2 - 1 = 0$, then $l_1(x)$ satisfies the equation $l_1(x) + 2x_2 = 0$, and it follows that $\Phi_0(x)$ is the function $x_1^2 + x_2^2 - 2x_2(x_2 - 1)$.

Therefore the device shown by equation (3.6.6) is useful, and we let $\Phi_1(x)$ be the function

$$\Phi_1(x) = F(x) + \sum_{i=1}^{\gamma} l_i(x)c_i(x) + \rho \sum_{i=1}^{\gamma} [c_i(x)]^2, \tag{3.6.15}$$

where ρ is a positive *constant*. Fletcher (1970a) proves that if the matrix $N(x^*)$ has rank γ, and if certain derivatives exist, then usually there exists a number, $\bar{\rho}$ say, such that for $\rho > \bar{\rho}$, x^* is a local minimum of $\Phi_1(x)$. Therefore if at the beginning of the calculation we set ρ to any number that exceeds $\bar{\rho}$, then usually x^* can be calculated by applying an algorithm for unconstrained minimization to the function (3.6.15) once only.

To fix the value of ρ, it is helpful to consider the function (3.6.15), when $F(x)$ is a positive-definite quadratic function, say

$$F(x) = \bar{g}^T x + \tfrac{1}{2} x^T G x, \tag{3.6.16}$$

and when the functions $c_i(x)$ $(i = 1, 2, \ldots, \gamma)$ are linear, say

$$c_i(x) = \sum_{j=1}^{n} N_{ji} x_j + d_i. \tag{3.6.17}$$

In this case $\Phi_1(x)$ is the quadratic function

$$\Phi_1(x) = \bar{g}^T x + \tfrac{1}{2} x^T G x + l(x)^T (N^T x + d) + \rho \| N^T x + d \|^2, \tag{3.6.18}$$

where $l(x)$ satisfies equation (3.6.13), so it is the vector

$$l(x) = -N^+(Gx + \bar{g}), \tag{3.6.19}$$

N^+ being the pseudo inverse of N. It follows that the value of ρ has to be so large that the matrix

$$G(\Phi_1) = G - G(NN^+) - (NN^+)G + 2\rho NN^T \tag{3.6.20}$$

is positive definite. It can be shown that this matrix is positive definite if and only if ρ exceeds the largest eigenvalue of the matrix $\tfrac{1}{2} N^+ G N^{+T}$. Therefore the quantity $\| N^+ G N^{+T} \|$ is a moderate value for ρ (Fletcher, 1970).

Now when solving problems with inequality constraints, one may guess which are the active constraints, and one may have to modify this guess during the calculation. Therefore it is unsatisfactory that the value $\rho = \| N^+ G N^{+T} \|_2$ depends on N. Fortunately we can avoid the dependence on N by using the function

$$\Phi_2(x) = F(x) + \sum_{i=1}^{\gamma} l_i(x)c_i(x) + \rho \| N^{+T}(x) c(x) \|_2^2 \tag{3.6.21}$$

in place of the function (3.6.15). Indeed we find that if equations (3.6.16) and (3.6.17) hold, then the second derivative matrix of $\Phi_2(x)$ is

$$G(\Phi_2) = G - G(NN^+) - (NN^+)G + 2\rho NN^+. \tag{3.6.22}$$

Therefore $G(\Phi_2)$ is positive definite if and only if ρ exceeds the greatest value of $\frac{1}{2}(u^T G u)/\|u\|^2$, where u is any vector in the column space of N. It follows that it is sufficient to let $\rho > \frac{1}{2}\|G\|_2$, so Fletcher recommends applying an algorithm for unconstrained minimization to the function (3.6.21), where $\rho = \|G\|_2$.

In practice the function $F(x)$ is seldom quadratic, so if the value $\rho = \|G\|_2$ is used, then, for want of anything better, $\|G\|_2$ has to be estimated at an initial point x that may be remote from x^*. Therefore it is possible that the value of ρ may have to be changed during the calculation. Fletcher and Lill (1970) report that for most of the test problems that they tried they did not have to reset ρ, and on the remaining test problems they had to reset ρ only once.

The main disadvantage of the algorithm is that the objective functions (3.6.14), (3.6.15) and (3.6.21) all depend on first derivatives, and therefore if the algorithm for unconstrained minimization requires exact first derivatives of the objective function, then second derivatives of $F(x)$ and $c_i(x)$ $(i = 1, 2, \ldots, \gamma)$ have to be calculated. Specifically the gradient of the function (3.6.21) is the vector

$$\nabla\Phi_2(x) = g - NN^+g - GN^{+T}c - c^T[\nabla N^+]g + 2\rho N^{+T}c + 2\rho c^T[\nabla N^+]N^{+T}c, \quad (3.6.23)$$

where g is the gradient of $F(x)$, where G is the second derivative matrix of $F(x)$, and where $[\nabla N^+]$ is a three suffix quantity.

We would prefer to avoid an exact calculation of $\nabla\Phi_2(x)$, so Fletcher and Lill experimented with a number of approximations to this vector. Fortunately they find that it seems to be adequate to omit the $[\nabla N^+]$ terms, and to replace G by a matrix B, that is obtained from calculated gradients of $F(x)$. Therefore they recommend the approximation

$$\nabla\Phi_2(x) \approx g - NN^+g - BN^{+T}c + 2\rho N^{+T}c, \quad (3.6.24)$$

which avoids the calculation of all second derivatives. They give a number of numerical examples that justify this approximation. Also they discuss the choice of algorithm for the unconstrained minimization.

In this chapter it has been shown that algorithms for unconstrained minimization are important to many methods for treating constraints on the variables. It seems likely that algorithms of the type described in Section 3.6 will in time supersede the penalty function methods of Section 3.5.

4. Second Derivative Methods

W. MURRAY
National Physical Laboratory

4.1 Introduction

If it is given that $F(x) \in C^2$ and that its Hessian matrix G can be computed for any x then there are algorithms which can utilize this knowledge. These algorithms are nearly all variants of Newton's method. In this chapter a number of the better-known algorithms are described together with a previously unpublished algorithm.

Naturally, even if G is available, we could use one of the many algorithms for which this knowledge is not required. An obvious reason for doing this is to alleviate the need to do the analytical differentiation and subsequent computer programming. This disadvantage can be partly answered by the availability of computer programs that will perform analytical differentiation. Although these programs may not produce an efficient code they are more reliable than programs based on differentiation by hand. The need for a computer program to evaluate G is an additional feature which does not arise in algorithms which do not require G. To justify the use of a second derivative method there must therefore be some advantages. One that can be appreciated immediately is that it is nearly always possible to check whether $\overset{*}{x}$ satisfies both the necessary and *sufficient* conditions for a strong local minimum. The main hope is that second derivative methods will prove more reliable and take significantly fewer iterations than alternative methods.

4.2 The method of steepest descent

This is one of the oldest methods for minimization which in its classical form does not require the Hessian matrix G. The purpose of describing this method is to demonstrate that most of the variants of Newton's method can be described as steepest descent procedures. The difference between the methods lies in the choice of norm used to define "steepest".

A single iteration of the classical method of steepest descent is given by, k*th iteration*

$$p^{(k)} = -g^{(k)},$$
$$x^{(k+1)} = x^{(k)} + \alpha^{(k)} p^{(k)},$$

where $\alpha^{(k)}$ is chosen to minimize $F(x^{(k)} + \alpha p^{(k)})$ with respect to α.

The significance of this choice of $p^{(k)}$ is that it minimizes the first variable term in the Taylor expansion of $F(x)$ for a given length of step. The Taylor expansion of $F(x)$ about $x^{(k)}$ for some change p is given by

$$F(x^{(k)} + p) = F(x^{(k)}) + p^T g^{(k)} + \ldots .$$

Consider the problem

$$P \qquad \underset{p}{\text{minimize}} \{p^T g^{(k)}\} \quad ,$$

subject to

$$\|p\| = \beta \quad ,$$

where β is an arbitrary positive scalar.

The direction $p^{(k)}$, the solution to P, is invariant with respect to changes in β. The solution to P for the norm

$$\|y\|^2 = y^T y, \tag{4.2.1}$$

is given by

$$p^{(k)} = -\beta g^{(k)} / \|g^{(k)}\|.$$

This is identical to the direction of search in the method of steepest descent. A point of significance is that regardless of the norm chosen there exists a value of β for which

$$F(x^{(k)} + p^{(k)}) < F(x^{(k)}),$$

provided $\|g^{(k)}\| \neq 0$.

There is a considerable amount of theoretical knowledge about the classical method of steepest descent. It is possible to prove convergence under weak assumptions on $F(x)$ and also to obtain bounds on the asymptotic rate of convergence. Unfortunately the bounds demonstrate that the method is likely to perform unsatisfactorily on the majority of problems. This slow rate of convergence persists even on certain quadratic functions. For any algorithm to perform well on a general nonlinear function it is a prerequisite that it performs well on a quadratic function.

Suppose we have a strictly convex quadratic function, that is

$$F(x) = \tfrac{1}{2} x^T G x + b^T x,$$

where G is positive definite.

The solution to P for the norm

$$\|y\|^2 = y^T G y, \tag{4.2.2}$$

is given by $p^{(k)}$ where

$$p^{(k)} = -\frac{\beta}{(g^{(k)T}G^{-1}g^{(k)})^{\frac{1}{2}}} G^{-1}g^{(k)}. \tag{4.2.3}$$

If $\alpha^{(k)}$ is chosen to minimize $F(x^{(k)}+\alpha p^{(k)})$ with respect to α then

$$x^{(k+1)} = x^{(k)}+\alpha^{(k)}p^{(k)} = \overset{*}{x},$$

where $F(\overset{*}{x})$ is the minimum value of $F(x)$. We can choose β so that $\alpha^{(k)} = 1$ and this gives

$$p^{(k)} = -G^{-1}g^{(k)}. \tag{4.2.4}$$

The superiority of this direction of search over that obtained using the norm (4.2.1) leads to the speculation that this $p^{(k)}$ may prove superior on a general non-linear function.

4.3 Newton's method

Since so many methods in numerical analysis are called Newton's method it becomes necessary to define exactly what we mean by this term. Suppose in an algorithm for a general function $F(x) \in C^2$ the direction of search is chosen to be that given by (4.2.3). The kth iteration is then given by

*k*th *iteration*

$$p^{(k)} = -G^{(k)-1}g^{(k)},$$
$$x^{(k+1)} = x^{(k)}+\alpha^{(k)}p^{(k)},$$

where $\alpha^{(k)}$ is chosen to minimize $F(x^{(k)}+\alpha p^{(k)})$ with respect to α.

This will be called Newton's method.

There are a number of observations that can be made. It is not a steepest-descent algorithm since $G^{(k)}$ is not in general positive definite so that

$$(y^T G^{(k)} y)^{\frac{1}{2}}$$

is not an admissible norm of y. It follows from this that $\alpha^{(k)}$ is not necessarily positive. Since $p^{(k)}$ is not defined when $G^{(k)}$ is singular, it is necessary to modify the algorithm. Numerical experience even on functions for which $G^{(k)}$ is non singular suggests that modification to the definition of $p^{(k)}$ is also necessary when $G^{(k)}$ is indefinite. There have been a number of suggestions to circumvent these difficulties. Some of these have been successful, but others, as will be shown, introduce certain unsatisfactory features.

4.4 Greenstadt's method

To improve the direction of search, when $G^{(k)}$ is indefinite, Greenstadt (1967) suggested the following variant of Newton's method.

Let $\lambda_j^{(k)}$ be the jth eigenvalue of $G^{(k)}$ and $v_j^{(k)}$ its corresponding eigenvector, with $v_j^{(k)T}v_j^{(k)} = 1$.

Define the $n \times n$ matrix $\bar{G}^{(k)}$ as

$$\bar{G}^{(k)} = \sum_{j=1}^{n} |\lambda_j^{(k)}| v_j^{(k)} v_j^{(k)T}.$$

The direction of search in the kth iteration of Newton's method is replaced by

$$p^{(k)} = -\bar{G}^{(k)-1} g^{(k)}.$$

This is still an inadequate definition of $p^{(k)}$ since $\bar{G}^{(k)}$ could be singular. To avoid this difficulty we redefine $\bar{G}^{(k)}$ as follows

$$\bar{G}^{(k)} = \sum_{j=1}^{n} \beta_j v_j^{(k)} v_j^{(k)T},$$

where

$$\beta_j = \max(|\lambda_j^{(k)}|, \delta).$$

The positive scalar δ is machine dependent. If a machine has a t bit word length then the usual choice for δ is

$$\delta = 2^{-t/2}.$$

The matrix $\bar{G}^{(k)}$ is positive definite so this variant of Newton's method is a steepest descent algorithm under the norm

$$\|y\|^2 = y^T \bar{G}^{(k)} y.$$

It follows from the definition of $\bar{G}^{(k)}$ that

$$\bar{G}^{(k)-1} = \sum_{j=1}^{n} \beta_j^{-1} v_j^{(k)} v_j^{(k)T},$$

consequently the vector $p^{(k)}$ can be calculated in only a further $O(n^2)$ operations once the $\lambda_j^{(k)}$ and $v_j^{(k)}$ are known. The method is satisfactory although the amount of work necessary to compute the eigenvalues and eigenvectors is high. This leads us to suggest the following scheme that avoids unnecessary calculation of the eigenvalues and eigenvectors. If $G^{(k)}$ is positive definite, then the only work necessary to determine $p^{(k)}$ is to solve the equations

$$G^{(k)} p^{(k)} = -g^{(k)}.$$

The most effective way of doing this (Wilkinson, 1965) is to factorize $G^{(k)}$, by the method of Cholesky, into the form

$$G^{(k)} = LDL^T,$$

where L is a lower-triangular matrix with unit-diagonal elements and D is a diagonal matrix. This method requires $\frac{1}{6}n^3 + O(n^2)$ multiplications compared with a conservative estimate of $3n^3$ multiplications for the eigenvector analysis. If $G^{(k)}$ is not positive definite then this will be revealed in the factorization by the occurrence of a non-positive diagonal in D.

We can then abandon the process and perform an eigenvector analysis. Should this occur at every iteration we will at most have increased the number of computer operations required by 6%. If for half the iterations $G^{(k)}$ is positive definite then we will have reduced the number of operations required by nearly a half. A more usual figure is that $G^{(k)}$ is positive definite 95% of the time so that the saving in this case would be more than a 90% reduction.

4.5 The Marquardt–Levenberg method

This method is an adaptation of the Marquardt–Levenberg algorithm for the solution of non-linear least-squares problems described by Powell in Chapter 3. In this method the direction of search is given by

$$p^{(k)} = -\bar{G}^{(k)-1}g^{(k)},$$

where

$$\bar{G}^{(k)} = (G^{(k)} + \beta^{(k)}Q^{(k)}),$$

$\beta^{(k)}$ being a non-negative scalar and $Q^{(k)}$ some specified matrix which is at least positive semidefinite. Among the suggestions for $Q^{(k)}$ have been

$$Q^{(k)} = I, \qquad \text{the unit matrix}$$

and

$Q^{(k)} = D^{(k)}$, a diagonal matrix whose elements are the absolute value of the diagonal elements of $G^{(k)}$. The scalar $\alpha^{(k)}$ is taken to be unity whilst $\beta^{(k)}$ is chosen so that $F(x^{(k+1)}) < F(x^{(k)})$ and $\bar{G}^{(k)}$ is a positive-definite matrix. The method generates the steepest descent step under the norm

$$\|y\|^2 = y^T\bar{G}^{(k)}y.$$

A disadvantage of the method is that a suitable $\beta^{(k)}$ is not known initially and with each estimate of $\beta^{(k)}$ a new set of linear equations must be solved in order to determine the corresponding $p^{(k)}$. Bard (1970), when solving the non-linear least-squares problem by this method with $Q^{(k)} = I$, suggested a means of avoiding this difficulty. Let λ_j be the jth eigenvalue of $G^{(k)}$ and v_j its corresponding eigenvector, with $v_j^T v_j = 1$, then

$$G^{(k)-1} = \sum_{j=1}^{n} \lambda_j^{-1} v_j v_j^T,$$

and

$$\bar{G}^{(k)-1} = \sum_{j=1}^{n} (\lambda_j + \beta^{(k)})^{-1} v_j v_j^T.$$

Hence once the λ_j's and v_j's are known each of the trial $p^{(k)}$'s can be determined in $O(n^2)$ operations.

Examination of the relative amounts of work required for solving sets of linear equations and computing eigensystems, shows that this scheme would only be worthwhile if 20 or more trial $\beta^{(k)}$'s were required. Each trial $\beta^{(k)}$ has an associated function evaluation, the test of a satisfactory $\beta^{(k)}$ being that this function value is less than the current lowest function value. If the algorithm required 20 function evaluations per iteration it would be judged very unsatisfactory. A further disadvantage is that storage of the eigenvectors requires twice the number of locations as that required for the Cholesky factors.

4.6 Cholesky's method for factorizing a positive-definite symmetric matrix

The remaining three methods that will be described are all based on utilizing the Cholesky factors of $G^{(k)}$. The first two methods described are numerically unstable and are not considered in any more detail than that necessary to indicate why. It is necessary for an understanding of the methods to be familiar with Cholesky factorization of a matrix so this is now described.

If A is an $n \times n$ positive-definite symmetric matrix it can be factorized in the form

$$A = LDL^T, \tag{4.6.1}$$

where L is a lower-triangular matrix with unit diagonal elements and D is a diagonal matrix whose diagonal elements are positive.

The factorization (4.6.1) is referred to as the Cholesky factorization of A and the matrices L and D as the Cholesky factors of A. The elements of L and D, which are unique, can be determined either column by column or row by row from equating elements in (4.6.1). Suppose the first $k-1$ columns of L and D have been determined, then the kth columns can be determined from the following relationships.

Let $l_{i,j}$ and $a_{i,j}$ denote the (i, j)th element of L and A respectively and d_j the jth diagonal element of D. Equating the (k, k)th element in (4.6.1) gives

$$\sum_{i=1}^{k} d_i l_{k,i}^2 = a_{k,k}, \tag{4.6.2}$$

which on rearranging becomes

$$d_k = a_{k,k} - \sum_{i=1}^{k-1} d_i l_{k,i}^2 \tag{4.6.3}$$

By equating further elements in the kth column of A we obtain

$$\sum_{i=1}^{k} d_i l_{j,i} l_{k,i} = a_{j,k}, \quad j = k+1, \ldots, n \tag{4.6.4}$$

which rearranges to give

$$l_{j,k} = \left(a_{j,k} - \sum_{i=1}^{k-1} d_i l_{j,i} l_{k,i} \right) \Big/ d_k, \quad j = k+1, \ldots, n. \tag{4.6.5}$$

It could happen that d_k is very small, but it follows from the relationship (4.6.2) that this does not result in a large element in $LD^{\frac{1}{2}}$ except possibly when A has a large *diagonal* element. The numerical stability of Cholesky's method is a direct result of the *a priori* bounds on the elements of $LD^{\frac{1}{2}}$ given by (4.6.2). If A is not a positive-definite matrix then the factorization (4.6.1) need not exist. If there is such a factorization then at least one of the elements of D is negative and (4.6.2) can no longer be used to give *a priori* bounds on the elements of LD.

4.7 The method of Fiacco and McCormick

A full description of this method is given in Fiacco and McCormick (1968). The first step of the kth iteration is to factorize $G^{(k)}$ into its Cholesky factors, that is

$$G^{(k)} = L^{(k)}D^{(k)}L^{(k)T}. \tag{4.7.1}$$

If the diagonal elements of $D^{(k)}$ are positive the direction of search, $p^{(k)}$, is determined as in Newton's method. Should any diagonal element be zero or negative, an alternative method for obtaining $p^{(k)}$ is given. This alternative, which involves the use of $L^{(k)}$ and $D^{(k)}$, cannot be relied upon (Wilkinson, 1965). The consequence is that $p^{(k)}$ may now be an arbitrary descent direction in no way related to $G^{(k)}$.

4.8 The method of Matthews and Davies

This method was first described by Matthews and Davies in 1969 and then in a slightly modified form in 1971. It was stated in the original version that the algorithm was based on two principles. Unfortunately neither of these principles was true and as a consequence the algorithm was numerically unstable. Although the offending principles have been removed in the later version and the algorithm slightly modified, it remains numerically unstable.

The first step of the kth iteration is to factorize the matrix $G^{(k)}$ into the form

$$G^{(k)} = L^{(k)}U^{(k)}, \tag{4.8.1}$$

where $L^{(k)}$ is a lower-triangular matrix with unit diagonal elements and $U^{(k)}$ is an upper-triangular matrix.

Matthews and Davies propose that this factorization should be performed by Gaussian elimination. This method is similar to Cholesky's method but takes twice as long and uses twice the storage. It is clear from assumptions they make about the factorization that in their implementation of Gaussian elimination they pivot down the diagonal. If $G^{(k)}$ is positive definite then the matrix $L^{(k)}$ is identical to that given by Cholesky's method and

$$U^{(k)} = D^{(k)}L^{(k)T}.$$

During the course of the elimination process the diagonal elements of $U^{(k)}$ are altered according to the rules,

$$\text{for } u_{i,i} \neq 0 \text{ set } u_{i,i} = |u_{i,i}|,$$
$$\text{for } u_{i,i} = 0 \text{ set } u_{i,i} = 1{\cdot}0.$$

This does not affect the process if $G^{(k)}$ is positive definite since all the diagonal elements of $U^{(k)}$ are then positive. Unfortunately if $G^{(k)}$ is not positive definite there is nothing in the modifications that will guarantee that the elements of $U^{(k)}$ and $L^{(k)}$ do not become arbitrarily large. The consequences are identical to those of Fiacco and McCormick's method.

4.9 A numerically stable modified Newton method based on Cholesky factorization

Previously a matrix, say A, has been factorized into the form

$$A = LDL^T. \tag{4.9.1}$$

Obviously if A is positive-definite we could have used the factorization

$$A = \bar{L}\bar{L}^T, \tag{4.9.2}$$

where $\bar{L}$ is a lower-triangular matrix. The relationship between $\bar{L}$ and L is simply

$$\bar{L} = LD^{\frac{1}{2}},$$

where

$$(D^{\frac{1}{2}})^2 = D.$$

Normally the factorization (4.9.1) is preferred to (4.9.2) because it avoids the need to form the square roots of the elements of D. In the modifications to the factorization process we shall describe, the square roots occur irrespective of which factorization is used. Consequently we use the factorization (4.9.2) since this simplifies the description. The equivalent relations to (4.6.2), (4.6.3) and (4.6.5) are

$$\sum_{i=1}^{k} l_{k,i}^2 = a_{k,k} \tag{4.9.3}$$

$$l_{k,k} = \left(a_{k,k} - \sum_{i=1}^{k-1} l_{k,i}^2\right)^{\frac{1}{2}} \tag{4.9.4}$$

and

$$l_{j,k} = \left(a_{j,k} - \sum_{i=1}^{k-1} l_{j,i} l_{k,i}\right) \Big/ l_{k,k}, \quad j = k+1, \ldots, n. \tag{4.9.5}$$

respectively.

In the proposed algorithm an initial direction of search in the kth iteration is found by solving the equations

$$\bar{L}^{(k)}\bar{L}^{(k)T}p^{(k)} = -g^{(k)}, \tag{4.9.6}$$

where $\bar{L}^{(k)}$ is a lower-triangular matrix.

The matrix $\bar{L}^{(k)}$ is determined by applying a variant of Cholesky's method to the matrix $G^{(k)}$. The resulting factorization will not necessarily be that of $G^{(k)}$ but of some other matrix say $\bar{G}^{(k)}$. The deviations from Cholesky's method occur when tests carried out during the process of factorizing $G^{(k)}$ indicate that $G^{(k)}$ is not positive-definite. The objective of the changes is to ensure that $\bar{G}^{(k)}$ differs from $G^{(k)}$ in some minimum way.

Suppose the modified factorization procedure is applied to a symmetric but indefinite matrix A, whose smallest eigenvalue is λ. Let the resulting factor computed be $\bar{L}$. A prerequisite for A to differ from $\bar{L}\bar{L}^T$ in some minimum way is for the quantity

$$\|A - \bar{L}\bar{L}^T\|$$

to be a continuous function of λ which is zero when λ is zero. In practice it is not possible to achieve this ideal since there is a need to bound the condition number of $\bar{G}^{(k)}$ in order to avoid numerical difficulties in evaluating $p^{(k)}$. Consequently in our modification

$$\|G^{(k)} - \bar{G}^{(k)}\|$$

tends uniformly to zero as λ tends to δ, a small positive scalar related to the word length of the computer employed.

The important property that is lost when applying Cholesky's method to a matrix which is not positive-definite (in practice this is also lost when the matrix is positive-definite but very close to a singular matrix) is the *a priori* bound on the elements of $\bar{L}$ given by (4.9.3). In the modification we propose this property is not required since the procedure acts directly to limit the size of the elements. It is clear from (4.9.5) that the off-diagonal elements of $\bar{L}$ could always be reduced, if they proved too large, by increasing the diagonal element of $\bar{L}$.

If the factorization is performed row by row then it is not possible to bound the change necessary in the jth diagonal element in order to bound the remaining elements in the jth row. It will be shown that it is possible to bound the change in the diagonal elements if the factorization is performed column by column.

To simplify the description of the modification, the superfix k will be temporarily dropped. What will be described is how the ith column of $\bar{L}$ is determined given that the first $i-1$ columns are known and that

$$|l_{r,s}| \leqslant \beta \quad r = 2, \ldots, n, \quad s = 1, \ldots, r-1, \tag{4.9.7}$$

where β is some preassigned scalar.

Define

$$\hat{l}_i = \max\left\{\delta, \left|G_{i,i} - \sum_{r=1}^{i-1} l_{i,r}^2\right|^{\frac{1}{2}}\right\} \tag{4.9.8}$$

and

$$\hat{l}_j = \left(G_{i,j} - \sum_{r=1}^{i-1} l_{i,r} l_{j,r}\right) \Big/ l_i, \quad j = i+1, \ldots, n \quad (4.9.9)$$

where δ is some small positive scalar dependent on the word-length of the computer being employed. A suitable choice of this parameter on a computer with a t bit word-length is

$$\delta = 2^{-t/2}.$$

If $\theta \leqslant \beta$, where

$$\theta = \max \{|\hat{l}_j| \mid j = i+1, \ldots, n\},$$

then

$$l_{j,i} = \hat{l}_j, \quad j = i, i+1, \ldots, n \quad (4.9.10)$$

If $\theta > \beta$, then

$$l_{i,i} = \theta \hat{l}_i / \beta \quad (4.9.11)$$

and

$$l_{j,i} = \beta \hat{l}_j / \theta, \quad = i+1, \ldots, n. \quad (4.9.12)$$

It is clear no matter which of the two definitions is used that

$$|l_{j,i}| \leqslant \beta \quad j = i+1, \ldots, n.$$

We shall now show that the diagonal elements are bounded. To do this it is first necessary to obtain bounds on the elements $|\hat{l}_j$, $j = i, \ldots, n$. It follows from (4.9.8) that

$$\hat{l}_i \leqslant \left(\left|G_{i,i}\right| + \sum_{r=1}^{i-1} l_{i,r}^2\right)^{\frac{1}{2}} + \delta,$$

hence

$$\hat{l}_i \leqslant \left(\left|G_{i,i}\right| + \sum_{r=1}^{i-1} \beta^2\right)^{\frac{1}{2}} + \delta,$$

which implies

$$\hat{l}_i \leqslant (|G_{i,i}| + (i-1)\beta^2)^{\frac{1}{2}} + \delta. \quad (4.9.13)$$

Similarly it follows from (4.9.9) that

$$|\hat{l}_j| \leqslant \theta \leqslant (\xi_i + (i-1)\beta^2) | \hat{l}_i, \quad j = i+1, \ldots, n, \quad (4.9.14)$$

where

$$\xi_i = \max \{|G_{i,j}| \mid j = i+1, \ldots, n\}.$$

If $\theta \leqslant \beta$ then $l_{i,i} = \hat{l}_i$ and $l_{i,i}$ is bounded by the inequality (4.9.13) else if $\theta > \beta$

$$l_{i,i} \leqslant (\xi_i + (i-1)\beta^2)/\beta \quad ,$$

which implies

$$l_{i,i} \leqslant \xi_i/\beta + (i-1)\beta. \quad (4.9.15)$$

This completes the proof that the diagonal elements of $\bar{L}$ are bounded.

The choice of the parameter β is affected by two criteria. The first criterion is that β should be large enough so that if G is sufficiently positive-definite $\bar{G} = G$, and the second is the wish to minimize the bound (4.9.15) for all i.

It follows from (4.9.3) that if

$$\beta^2 \geqslant \max_i \{|G_{i,i}|\} = \gamma.$$

the first criterion is satisfied. The bound on the diagonal elements will be minimized if

$$\beta^2 = \xi/\gamma,$$

where

$$\xi = \max_i \{\xi_i\}.$$

Therefore β is defined to be

$$\beta = \max \{\gamma^{\frac{1}{2}}, (\xi/n)^{\frac{1}{2}}\}. \tag{4.9.16}$$

It follows from (4.9.15) and (4.9.16) that

$$\delta \leqslant l_{i,i} \leqslant 2n\beta, \quad i = 1, \ldots, n. \tag{4.9.17}$$

It is important to realize that the definition of the off-diagonal elements given by (4.9.12) is identical to that which would have resulted if, in applying Cholesky's method, the diagonal elements had been given by (4.9.11). The lower-triangular matrix obtained by this procedure is therefore identical to that which would have been obtained by applying Cholesky's method to the matrix

$$\bar{G} = G + D\ , \tag{4.9.18}$$

where D is a diagonal matrix.

Clearly the elements of D are bounded. It follows from (4.9.18) that the ith element of D, d_i, is given by

$$d_i = \sum_{j=1}^{i} l_{i,j} - G_{i,i}. \tag{4.9.19}$$

Note that if

$$l_{i,i} = \left(G_{i,i} - \sum_{j=1}^{i-1} l_{i,j}\right)^{\frac{1}{2}}$$

then

$$d_i = 0.$$

The number of operations taken by the procedure is $\frac{1}{6}n^2 + O(n^2)$ and since $\bar{L}$ can be overwritten on G the additional storage required is negligible.

The superfix k will now be reintroduced. The direction of search in the kth iteration is found by solving

$$\bar{L}^{(k)}\bar{L}^{(k)}p^{(k)} = -g^{(k)},$$

and

$$x^{(k+1)} = x^{(k)} + \alpha^{(k)} p^{(k)}.$$

Since $\bar{G}^{(k)}$ is positive-definite $p^{(k)}$ is the steepest-descent direction under the norm

$$\|y\|^2 = y^T(G^{(k)} + D^{(k)})y.$$

An alternative procedure which avoids the need to determine $\alpha^{(k)}$ is to define $p^{(k)}$ to be the steepest-descent direction under the norm

$$\|y\|^2 = y^T(\bar{L}^{(k)} + \lambda \hat{L})(\bar{L}^{(k)} + \lambda \hat{L})^T y,$$

where $\hat{L}$ is some specified lower-triangular matrix and λ is a non-negative scalar chosen so that

$$F^{(k+1)} < F^{(k)}, \tag{4.9.20}$$

if

$$x^{(k+1)} = x^{(k)} + p^{(k)}.$$

Possible choices of $\hat{L}$ are

(i) $L^{(k-1)}$.
(ii) I.
(iii) $D^{(k)}$.
(iv) $\hat{D}$.

where $\hat{D}$ is a diagonal matrix whose ith diagonal element is $\bar{l}_{i,i}^{(k)}$. The first choice has the effect of biasing $p^{(k)}$ towards the previous direction of search. It has the disadvantage of the need to store an additional lower-triangular matrix. The remaining three choices will all satisfy (4.9.20) if λ is chosen large enough. This procedure is very similar to the Marquardt–Levenberg method except it has the advantages of avoiding the need to repeatedly solve a system of linear equations, uses less storage and enables a better choice to be made of the weighting matrix.

4.10 Saddle points

It is one of the advantages of second derivative methods that they are able to distinguish between most saddle points and local minima. The exceptions are those saddle points where the Hessian matrix has negative eigenvalues but ones so close to zero that the computer is unable to distinguish them from positive eigenvalues. Even in these circumstances it is very unlikely that the procedures described would converge to such a point.

As the methods have been described they would fail if $x^{(k)}$ was a saddle point, i.e.

$$g^{(k)} = 0,$$

and $G^{(k)}$ an indefinite matrix. A simple expediency in these circumstances is merely to make $x^{(k+1)}$ a perturbation of $x^{(k)}$ such that $g^{(k+1)} \neq 0$. It is possible to construct more elaborate techniques to deal with this difficulty which will in general be more efficient. We do not consider this worthwhile since if $x^{(k-1)}$ is in the neighbourhood of a saddle point it is a property of the methods that $p^{(k-1)}$ is chosen so as not to pass through the saddle point. Consequently the likelihood of $x^{(k)}$ being a saddle point is very small.

4.11 Nonlinear least squares

This problem has already been discussed in some detail in Chapter 3. The purpose of introducing it again is simply to comment on how the last method described can be applied to the closely related problem of non-linear least squares.

The problem now is to

$$\text{minimize}\ \{F(x) = f(x)^T f(x)\}$$

where $f(x)$ is an $m \times 1$ vector of nonlinear functions. The gradient and Hessian matrix of $F(x)$ are given by

$$g = 2J^T\ (x)$$

and

$$G = 2J^TJ + 2\sum_{i=1}^{m} f_i(x)G_i$$

where J is the Jacobian matrix of $f(x)$

and G_i is the Hessian matrix of $f_i(x)$.

The methods described in Chapter 3 are based on forming the matrix $J^{(k)T}J^{(k)}$. In studying the case when $f_i(x)$ are linear functions, Businger and Golub (1965) have shown that merely forming $J^{(k)T}J^{(k)}$ results in an unnecessary loss of accuracy. This is still true for the non-linear case. The way of overcoming this loss suggested by Businger and Golub was to factorize $J^{(k)}$ directly by the repeated multiplication of Householder matrices, that is

$$J^{(k)T} = [L^{(k)}|0]Q^{(k)T}$$

where $L^{(k)}$ is a lower-triangular matrix

and $Q^{(k)}$ is an orthonormal matrix.

If the direction of search is given by solving the normal equations, then

$$L^{(k)T}p^{(k)} = b^{(k)},$$

where $b^{(k)}$ is the $n \times 1$ vector consisting of the first n elements of the vector

$$Q^{(k)T}f^{(k)},$$

where

$$f^{(k)} = f(x_k).$$

It is not necessary to store the matrix $Q^{(k)}$ since the vector $b^{(k)}$ can be formed during the factorization process.

Should a Marquardt–Levenberg type algorithm be required then $p^{(k)}$ can be defined to be the solution to the equations

$$(L^{(k)}+\lambda\hat{L})(L^{(k)}+\lambda\hat{L}^T)p^{(k)} = -J^{(k)T}g^{(k)}$$

which is identical to the technique described in § 4.9.

4.12 Comments and observations

The linear search in a second derivative method does not play such a critical role as a linear search does in most first derivative methods. In the methods that require a linear search a step that merely reduces the function will suffice. Another feature is that the initial predicted step to the minimum along a search direction is much more accurate than that given by first derivative methods. These two features result in the number of function evaluations per iteration being smaller for second derivative methods.

A question still to be answered is when should a second derivative method be used. This will depend on the following factors, most of which are unknown

- T_1 time to evaluate $F(x)$,
- T_2 time to evaluate g,
- T_3 time to evaluate G,
- T_4 time to execute one step of the algorithm knowing the function and its derivatives.

The ratio r of the total time to minimize a function compared with a first derivative method is given by

$$r = \frac{n_1 n_3 T_3 + n_2 n_3 (T_1 + T_3) + n_3 T_4}{n_4 n_5 (T_1 + T_3) + n_5 T_5} \tag{4.12.1}$$

where T_5 is the time to execute one step of a first derivative method given the function and gradient. Define

- n_3 number of iterations taken by second derivative method,
- n_5 number of iterations taken by first derivative method,
- n_2 average number of functions evaluations per iteration for second derivative method,
- n_4 average number of function evaluations per iteration for first derivative method,
- n_1 frequency the Hessian matrix is evaluated.

As the algorithms have been described $n_1 = 1$. If the initial estimate to the solution is poor then the expected savings in a reduced number of iterations may not be very great. To offset this the Hessian matrix need not be evaluated at each iteration. The frequency of evaluation can be related to the lack of progress using the old Hessian matrix and the size of the gradient. If the method described in § 4.9 is used and compared with a rank 2 quasi-

Newton algorithm

$$T_4/T_5 \doteq 1/6n^3/3n^2 \doteq n/18,$$

for large n. For $n < 20$, $T_4 \doteq T_5$. If we assume that

$$n_3/n_5 \doteq 1/3,$$

then

$$r \doteq \frac{n_1 T_3 + n_2(T_1+T_2)+T_3}{3n_4(T_1+T_2)+3T_5}.$$

We should use the second derivative method if

$$r < 1,$$

that is

$$T_3 < \frac{1}{n_1}((3n_4 - n_2)(T_1+T_2)+3T_5 - T_4). \qquad (4.12.2)$$

Assuming $n_1 = 1/3$ and $n_2 = 3$, $n_4 = 6$ and $n = 60$ so that the execution times cancel then (4.12.2) becomes

$$T_3 < 45(T_1+T_2)$$

Obviously this is a very rough guide but it does indicate that if second derivatives are available they almost certainly should be used except when their computation time is extremely large compared with that for the function and gradient.

5. Conjugate Direction Methods

R. FLETCHER

Atomic Energy Research Establishment, Harwell

5.1 Introduction

The motivation of many methods for minimization has been to set up a scheme of calculation which would enable the minimum of a quadratic function to be found in a finite number of iterations, and yet which could be applied iteratively to general functions. In doing this, other criteria have to be borne in mind. For instance, in the interests of efficiency, it is important that the subroutine to evaluate the function (and gradient where applicable) should not be called an unnecessarily large number of times. Now $\sim \frac{1}{2}n^2$ parameters are needed to determine a quadratic function. Hence methods which do not compute derivatives should minimize a quadratic function in $O(n^2)$ evaluations of the function, and methods which additionally compute first derivatives should require $O(n)$ evaluations of the function plus gradient vector. Similarly given second derivatives $O(1)$ evaluations should be necessary. There is a similar limit to the number of computer housekeeping operations which should be used. Given second derivatives, then $O(n^3)$ operations are required to minimize the quadratic function because of the need to solve a system of linear equations. Assuming that each iteration is a line search, this means that methods which only calculate function values should use $O(n)$ operations per iteration in order to be efficient in this respect, and methods which calculate first derivatives should use $O(n^2)$ operations per iteration. Finally, and of most importance the methods must be efficient in practice when applied to the minimization of general functions.

It will be assumed in this chapter that the matrix of second derivatives G of the function is not available, so the aim of setting up a scheme of calculation which satisfies the conditions of the previous paragraph must be done indirectly. The so-called *conjugate direction methods* are a whole class of methods which meet the objectives which have been set out. The following definition is fundamental.

Definition (1)

A set of vectors $p^{(1)}, p^{(2)}, \ldots, p^{(n)}$, where $p^{(i)} \neq 0$ are *conjugate* with regard to a given positive definite matrix G, if

$$p^{(i)T} G p^{(j)} = 0 \quad \text{for all } i \neq j. \tag{5.1.1}$$

This definition also implies that the vectors $\{p^{(i)}\}$ are mutually independent. The theory of conjugate direction methods relates almost entirely to quadratic functions, whence the Hessian matrix is constant. This assumption will be implicit in most of what follows except when discussing practical experience on general functions or in comments about convergence results which have been proved. The methods are all iterative and each iteration is a line search sub-problem; that is given an approximation to the solution $x^{(k)}$ and a search direction $p^{(k)}$, then $x^{(k+1)}$ is chosen so that

$$x^{(k+1)} = x^{(k)} + \alpha^{(k)} p^{(k)}, \tag{5.1.2}$$

where $\alpha^{(k)}$ is chosen to minimize $F(x^{(k)} + \alpha p^{(k)})$ with respect to α. There is no difficulty in doing this for a quadratic function, but such a process requires an infinite iteration when applied to general functions. Consequently most subroutines approximate the solution to (5.1.2) in a finite number of evaluations of the function (and possibly derivatives). For reasons of efficiency it is usual to try to evaluate the function as rarely as possible, and to accept quite crude estimates of the optimum. However, for the termination results on quadratic functions to be valid for any particular subroutine, that subroutine must have a line search which itself terminates in a finite number of function evaluations for a quadratic function. This would not be true for instance if Fibonnaci search was used.

The significance of definition (1) is that in conjugate direction methods the positive definite matrix G plays the role of the Hessian of the quadratic function and the vectors $p^{(i)}$ are the search directions. A theorem will now be proved which states essentially that

$$\text{conjugacy} + \text{line search} = \text{termination}$$

more formally this can be stated as follows.

Theorem 1

If each iteration is the line search (5.1.2) and the search directions $p^{(1)}, p^{(2)}, \ldots, p^{(n)}$ satisfy (5.1.1) with regard to a quadratic function with positive-definite Hessian G, then the minimum is found in at most n iterations, and moreover, $x^{(i+1)}$ is the minimum point in the subspace generated by the initial approximation $x^{(1)}$ and the directions $p^{(1)}, p^{(2)}, \ldots, p^{(i)}$.

Proof. Both results are proved by establishing the equations

$$g^{(i+1)T} p^{(j)} = 0, \quad i = 1, 2. \ldots, i, \tag{5.1.3}$$

for all $i \leqslant n$. By virtue of the line search, there follows

$$g^{(i+1)T}p^{(i)} = 0, \quad \text{for all } i \leqslant n. \tag{5.1.4}$$

Writing

$$g^{(i+1)T}p^{(j)} = g^{(j+1)T}p^{(j)} + (y^{(j+1)} + y^{(j+2)} + \ldots + y^{(i)})^T p^{(j)} \tag{5.1.5}$$

and using

$$y^{(j)} = Gs^{(j)} = \alpha^{(j)}Gp^{(j)}, \tag{5.1.6}$$

where $y^{(j)} = g^{(j+1)} - g^{(j)}$ and $s^{(j)} = x^{(j+1)} - x^{(j)}$, equations (5.1.3) are established by virtue of (5.1.4) and (5.1.1). QED.

Most of the conjugate direction methods which will be described rely upon this theorem, the differences between the methods being both in the ingenuity by which conjugate directions are generated without explicit knowledge of the Hessian, and in the appropriateness of the iterative process thus derived to deal with general functions. As usual the paper will be divided according to whether or not first derivatives are available explicitly, and the latter case is covered in section 5.2. The former problem is considered in section 5.3, where the most important method of conjugate gradients, and its variations, are discussed. In section 5.4 a number of other first derivative methods are introduced and in particular it is shown how conjugate gradient methods are related by virtue of their projection properties to the well-known quasi-Newton methods.

5.2 Minimization without derivatives

In this section the possibilities are considered for minimizing functions for which it is not convenient to write down explicit expressions from which first derivatives can be calculated. To make any progress towards a conjugate direction method it is necessary to have some property of quadratic functions which enables conjugacy to be obtained without using any derivatives of the function. Such a property is the "parallel subspace property" which can be stated in the form of a theorem, as follows:

Theorem 2

Consider two parallel subspaces S_1 and S_2, generated by independent directions $p^{(1)}, p^{(2)}, \ldots, p^{(k)}$, where $k < n$, from the points x and v respectively. (That is $S_1 = \{z | z = x + \sum_{i=1}^{k} \alpha_i p^{(i)} \forall \alpha_i\}$, S_2 similarly, and $S_1 \not\equiv S_2$.) Denote the points which minimize the quadratic function for $x \in S_1$ and $x \in S_2$ as $z^{(1)}$ and $z^{(2)}$ respectively. Then $z^{(2)} - z^{(1)}$ is conjugate to $p^{(1)}, p^{(2)}, \ldots, p^{(k)}$.

Proof The situation is illustrated in Fig. 1.

To prove the result is very simple, because by definition of a minimum

$$g(z^{(2)})^T p^{(i)} = g(z^{(1)})^T p^{(i)} = 0, \quad i = 1, 2, \ldots, k. \tag{5.2.1}$$

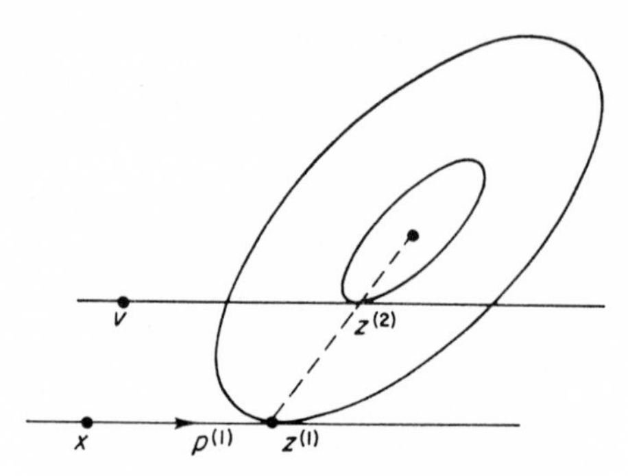

FIG. 5.1. The "parallel subspace" property

and hence

$$y^T p^{(i)} = 0, \tag{5.2.2}$$

where $y = g(z^{(2)}) - g(z^{(1)})$. By virtue of (5.1.6), $y = G(z^{(2)} - z^{(1)})$ from which

$$(z^{(2)} - z^{(1)})^T G p^{(i)} = 0, \quad i = 1, 2, \ldots, k \tag{5.2.3}$$

follows, proving the theorem. QED.

A method based on using this property to obtain conjugacy was suggested by Smith (1962). It assumes that a line search subroutine is available which uses only function values, and that, in addition to the initial approximation $x^{(1)}$, the user can supply independent directions $d^{(1)}, d^{(2)}, \ldots, d^{(n)}$ and constants $\beta_1, \beta_2, \ldots, \beta_n$, ($\beta_i > 0$), chosen so that, roughly speaking, $\beta_i d^{(i)}$ represents a moderate change in the variables. The algorithm as it would apply to a quadratic function can be stated as follows:

(i) Define $p^{(1)} = d^{(1)}$ and let $x^{(2)} = x^{(1)} + \alpha p^{(1)}$ be found by a line search along $p^{(1)}$.

(ii) For $i = 2, 3, \ldots, n$ note that $x^{(i)}$ is the minimum point in the subspace generated by $x^{(1)}$ and directions $p^{(1)}, p^{(2)}, \ldots, p^{(i-1)}$, and carry out the operations (a)–(c) for each i in turn.

(a) Displace $x^{(i)}$ to a point $v = x^{(i)} + \beta_i d^{(i)}$ which can be considered to be an arbitrary point in a parallel subspace.

(b) Displace v by making successive line searches along directions $p^{(1)}, p^{(2)}, \ldots, p^{(i-1)}$ in turn, giving a point $z = v + \sum_{j=1}^{i-1} \alpha_j p^{(j)}$ which is the minimum point in the parallel subspace by virtue of Theorem 1.

(c) Define $p^{(i)} = z - x^{(i)}$ and calculate $x^{(i+1)} = x^{(i)} + \alpha_i p^{(i)}$ by a line search along $p^{(i)}$. Note that $p^{(i)}$ is a new conjugate direction by virtue of Theorem 2, and hence that $x^{(i+1)}$ is the minimum point in the subspace generated by $x^{(1)}$ and the directions $p^{(1)}, p^{(2)}, \ldots, p^{(i)}$.

When step (ii) above has been completed, $x^{(n+1)}$ has essentially been found by minimizing from $x^{(1)}$ along a sequence of conjugate directions $p^{(1)}, p^{(2)}, \ldots, p^{(n)}$, and hence is the required minimum of the quadratic function. It will be noticed that this has been achieved by using $\frac{1}{2}n(n+1)$ line searches, and that the additional housekeeping is $O(n)$ per line search. Thus the method is efficient in these respects.

Unfortunately difficulties arise when applying the method iteratively to general functions of many variables. The obvious approach, as suggested by Smith, is merely to repeat the whole cycle described above, in an iterative manner. Practical experience (Fletcher, 1965) suggests that unless n is small ($\leqslant 4$ say) then this method is inferior to other methods, and the situation gets worse as n increases. Various modifications have been tried, for instance replacing step (ii, (a)) by a line search along $d^{(i)}$ so as to remove the need to choose β_i, and also orthogonalizing the $p^{(i)}$ after each cycle so as to provide new orthogonal directions for the next cycle (Fletcher, 1965), but neither has improved performance.

On examining the detailed progress of Smith's method, it is clear that the poor progress is caused by the fact that the directions of search receive unequal treatment in the number of times when they are used. In particular $p^{(1)}$ is used n times per cycle, whereas $p^{(n)}$ is used once. For general functions, because x is distant from the minimum in the direction $d^{(n)}$, because the function varies non-quadratically along $d^{(n)}$, and because a change in the direction $d^{(n)}$ is not made until the final step (ii) is carried out, many of the searches along $p^{(1)}$, for instance, are of little value. Powell (1964) made a significant advance in suggesting how searches could be incorporated into Smith's method, so as to treat all directions equally, whilst retaining the property of quadratic termination. In the context of Smith's method as described above, the modification consists of replacing step (ii, (a)), which is an arbitrary displacement, by a sequence of line searches along $d^{(i)}, d^{(i+1)}, \ldots, d^{(n)}$ in turn, all of which, excepting $d^{(i)}$, would not otherwise be introduced on that particular step (ii). As the displacement of step (ii, (a)) is arbitrary, it is seen that the quadratic termination property as proved for Smith's method is still valid. Furthermore, it is possible to rewrite the iteration in a very simple form, by coalescing (ii, (a)) and (ii, (b)), with the consequence that step (ii) permits of being continued for $i > n$, if the function is not quadratic, as follows:

(i) Given independent directions $p^{(1)}, p^{(2)}, \ldots, p^{(n)}$, let $x^{(1)}$ be the minimum point along $p^{(n)}$, by using a line search if necessary.

(ii) Repeat (a)–(d) for $i = 1, 2, \ldots, \infty$.

  (a) Note that for $i \leqslant n$ the last i directions $p^{(n-i+1)}, \ldots, p^{(n)}$ have been replaced by conjugate directions.

(b) Find $z = x^{(i)} + \sum_{j=1}^{n} \alpha_j p^{(j)}$ by making optimum line searches along $p^{(1)}, p^{(2)}, \ldots, p^{(n)}$.
(c) For $j = 1, 2, \ldots, n-1$ set $p^{(j)} = p^{(j+1)}$.
(d) Define $p^{(n)} = z - x^{(i)}$ as a new conjugate direction and let $x^{(i+1)} = z + \alpha p^{(n)}$ be found by an optimum line search along $p^{(n)}$.

The iteration now converges in $\sim n^2$ line searches which is about twice as many as for Smith's method. However, the method now not only treats directions equally, but allows an interpretation in terms of "pseudo-conjugate directions" which are retained for general functions when $i > n$, and are modified from iteration to iteration. These features make the method much preferable to Smith's method for minimizing general functions.

Unfortunately it was found that in some problems, the set $\{p^{(i)}\}$ tended to become linearly dependent in such a way that the true minimum was not located. This led Powell to suggest a modification to his algorithm in which steps (ii, (c) and (d)) were omitted if they would cause the set $\{p^{(i)}\}$ to become more linearly dependent. This modified algorithm proved superior to the DSC direct search method, and much superior to Smith's method in a comparison given by Fletcher (1965), and for some time was considered to be the best method for minimization without derivatives. However, the method does seem to lead to slow convergence if n is at all large, and a practical upper limit seems to be about 10–20 variables. This slow convergence is probably due to the test for omitting (c) and (d) being over restrictive, and perhaps a further modification should be sought such that directions are accepted if the degree of independence is not less than some preset tolerance. Another apparently promising modification has been suggested by Zangwill (1967), in which (c) and (d) are always retained, and is such that independence is achieved by injecting new directions chosen from a given independent set $d^{(1)}, d^{(2)}, \ldots, d^{(n)}$ on a regular basis. Zangwill has proved that, given exact arithmetic, his version of Powell's method converges; assuming that the function is strictly convex, and that the line searches are carried out exactly. No estimate of the rate of convergence is given. Although these conditions are unrealistic in that most users minimize non-convex functions, and it is not possible in general to carry out an exact line search; the convergence result does give some confidence in the reliability of the method. However, some recent work by Rhead (1971) suggests that in practice Zangwill's modification is worse than Powell's method, and he suggests a reason why this might be so. It may well be that there is scope for further developments in this field.

5.3 Minimization with first derivatives—the method of conjugate gradients

In the rest of the paper it will be assumed that it is possible to write down explicit expressions not only for $F(x)$, but also for the first partial derivatives

$\partial F/\partial x_i$ for any x. Most of the methods are based upon the line search (2), and again it will be assumed that such an algorithm is available. When derivatives are available a wide variety of line search algorithms exist; for instance some may calculate derivatives during the search whilst others may not. Although this is an important practical point, the only theoretical consideration for quadratic functions is that the line search terminates. The difficult problem is again that of achieving conjugacy for quadratic functions without evaluating the Hessian matrix G, and this is much easier when derivatives are available. This is because the conjugacy conditions (5.1.1) can be replaced, using (5.1.6), by

$$p^{(i)T}y^{(j)} = 0, \tag{5.3.1}$$

showing that it is merely necessary to force an orthogonality condition in order to achieve conjugacy.

The most important of the conjugate direction methods using derivatives will now be described. It is known as the *method of conjugate gradients* and may be thought of as an association of conjugacy (and hence termination) properties with steepest descent properties; in fact it chooses the steepest descent direction subject to the conjugacy conditions (5.3.1) being satisfied. The method owes much to the pioneering work of Hestenes and Stiefel (1952) on linear systems, but a more simple presentation will be given here. The iteration is based on using line searches (5.1.2) and for a quadratic function the choice of $\{p^{(i)}\}$ can be stated very simply as

$$p^{(1)} = -g^{(1)} \tag{5.3.2}$$

and for $i = 2, 3, \ldots, n$.

$p^{(i)}$ = the component of $-g^{(i)}$ which is orthogonal to

$$y^{(1)}, y^{(2)}, \ldots, y^{(i-1)}. \tag{5.3.3}$$

The most convenient way of stating this (although not of implementing it in practice) is

$$p^{(i)} = -Q^{(i)}g^{(i)}, \quad i = 1, 2, \ldots, n, \tag{5.3.4}$$

where $Q^{(i)}$ is the symmetric orthogonal projection matrix which annihilates $y^{(1)}, y^{(2)}, \ldots, y^{(i-1)}$. $Q^{(i)}$ can be defined explicitly as

$$Q^{(i)} = I - Y^{(i-1)}(Y^{(i-1)T}Y^{(i-1)})^{-1}Y^{(i-1)T}, \tag{5.3.5}$$

where $Y^{(i-1)}$ is the $n \times (i-1)$ matrix whose columns are the vectors $y^{(1)}$, $y^{(2)}, \ldots, y^{(i-1)}$. It is clear in addition that

$$Q^{(1)} = I, \tag{5.3.6}$$

$$Q^{(n+1)} = 0, \tag{5.3.7}$$

$$Q^{(i)}y^{(j)} = 0, \quad \text{for all } j < i, \tag{5.3.8}$$

$$Q^{(i)}v = v, \quad \text{for all } v \text{ such that } v^Ty^{(j)} = 0,\ j = 1, \ldots, i-1. \tag{5.3.9}$$

Two properties of the algorithm can now be determined very simply. First of all, the choice of $\{p^{(i)}\}$ in equation (5.3.4) ensures that the conjugacy conditions (5.1.1), so Theorem 1 can be invoked to prove quadratic termination. A second result can be stated in the following lemma.

Lemma 1

$$g^{(i)T}g^{(j)} = 0, \quad \text{for all } j < i \leqslant n+1. \tag{5.3.10}$$

Proof. Equations (5.3.4) and (5.3.5) show that $p^{(j)}$ is a linear combination of $g^{(1)}, g^{(2)}, \ldots, g^{(j)}$

$$p^{(j)} = \mathscr{L}(g^{(1)}, g^{(2)}, \ldots, g^{(j)}), \tag{5.3.11}$$

hence that

$$g^{(j)} = \mathscr{L}(p^{(1)}, p^{(2)}, \ldots, p^{(j)}). \tag{5.3.12}$$

Equation (5.1.3) now proves $g^{(i)T}g^{(j)} = 0$. QED.

It is now possible to consider the practical implementation of the method of conjugate gradients, and in particular to show that $p^{(i)}$ can be calculated *merely as the sum of two vectors.* By the definition (5.3.5) of $Q^{(i)}$ it follows that there exist scalars β_{ik} and β'_{ik} such that

$$p^{(i)} = -Q^{(i)}g^{(i)} = \beta_{ii}g^{(i)} + \sum_{k=1}^{i-1} \beta_{ik}g^{(k)} \tag{5.3.13}$$

and hence, by virtue of (5.3.12), that

$$p^{(i)} = \beta_{ii}g^{(i)} + \sum_{k=1}^{i-1} \beta'_{ik}p^{(k)}. \tag{5.3.14}$$

Multiplying (5.3.14) on the left by $y^{(j)}$, where $j < i$, and using the condition (5.3.1) gives

$$0 = \beta_{ii}y^{(j)T}g^{(i)} + \beta'_{ij}y^{(j)T}p^{(j)}. \tag{5.3.15}$$

Finally, by definition of $y^{(j)}$, and using Lemma 1, equation (5.3.15) can be rearranged to give

$$\beta'_{ij} = 0, \quad \text{for all } j < i-1. \tag{5.3.16}$$

Thus the somewhat surprising and very convenient result is obtained that all but two of the scalars in the expansion (5.3.14) are zero, and hence that

$$p^{(i)} = \beta_{ii}g^{(i)} + \beta'_{i,\,i-1}p^{(i-1)}. \tag{5.3.17}$$

Because a linear search process is used, it is possible to remove a constant factor from $p^{(i)}$ without changing the algorithm. Hence the choice of direction can now be written in terms of a single new scalar β_i as

$$\bar{p}^{(i)} = -g^{(i)} + \beta_i\bar{p}^{(i-1)}, \tag{5.3.18}$$

where $\bar{p}^{(i)}$ is a multiple of $p^{(i)}$. The scalar β_i can be determined most conveniently from the conjugacy condition $\bar{p}^{(i)T}y^{(i-1)} = 0$, giving

$$\beta_i = y^{(i-1)T}g^{(i)}/y^{(i-1)T}\bar{p}^{(i-1)}. \tag{5.3.19}$$

Furthermore, equations (5.3.18) and (5.3.11) provide the identities

$$\bar{p}^{(i-1)} = -g^{(i-1)} + \beta_{i-1}\bar{p}^{(i-2)} = -g^{(i-1)} + \mathscr{L}(g^{(1)}, g^{(2)}, \ldots, g^{(i-2)}),$$

so using Lemma 1, equation (5.3.19) can be written equivalently as

$$\beta_i = g^{(i)T}g^{(i)}/g^{(i-1)T}g^{(i-1)}. \tag{5.3.20}$$

The theory which has been developed about the method of conjugate gradients has so far only been appropriate to the case when $F(x)$ is a quadratic function, and it is important to consider what is the best way of applying the method for general functions. The two main possibilities are:

(i) to repeat the n iterations described above in cyclic fashion, resetting $\bar{p}^{(i)}$ to $-g^{(i)}$ every nth iteration, as follows.

$$\bar{p}^{(cn+1)} = -g^{(cn+1)}, \qquad c = 0, 1, 2, \ldots, \tag{5.3.21}$$

$$\bar{p}^{(i)} = -g^{(i)} + \beta_i\bar{p}^{(i-1)}, \quad \text{otherwise.} \tag{5.3.22}$$

(ii) to continue the two term recurrence (5.3.18) indefinitely, that is to define

$$\bar{p}^{(1)} = -g^{(1)}, \tag{5.3.23}$$

$$\bar{p}^{(i)} = -g^{(i)} + \beta_i\bar{p}^{(i-1)} \quad \text{for all } i > 1. \tag{5.3.24}$$

This problem was considered by Fletcher and Reeves (1964) who first considered applying this simple formulation of conjugate gradients to general minimization problems. They found that resetting as in (i) above was much superior to (ii) in practice, and I have since come across no evidence which contradicts this point of view. There is also an intuitive reason why (i) is better. One justification for developing methods which have a termination property, is that a general function resembles a quadratic function in a region close to a local minimum, and hence that such methods should converge rapidly in this region. It is therefore reasonable to expect that if the method is applied to a function which is arbitrary for the first $K > 0$ iterations, and quadratic thereafter, then termination should still occur. This model is a better approximation to the behaviour of a general function. Alternative (i) terminates under these conditions (as also do many other minimization methods, such as the DFP method), whereas alternative (ii) does not.

Other possible alternatives are, for instance, to reset $p^{(i)}$ to $-g^{(i)}$ less frequently, say on every $2n$ or $3n$ iterations. This has been recommended when solving linear equations as a means of compensating for the effects of round-off error but this aspect should not be given much weight in the context of general minimization. In fact, Fletcher and Reeves (1964) reset every $n+1$ iterations, not every n, although such a small difference is not significant. On the other hand, experiments have been tried in which $p^{(i)}$

is reset to $-g^{(i)}$ more frequently than on every nth iteration. Although the effect of these alternatives is quite different on any one problem, I have seen no evidence which indicates that any choice other than (i) should be made in a general purpose algorithm.

Another choice which should be examined is whether (5.3.19) or (5.3.20) (or any other choice) should be used to determine β_i. Choice (5.3.20) has been recommended when solving linear equations. However, this evidence, based on round-off considerations, should again be discounted for general minimization because (5.3.19) and (5.3.20) are different even neglecting the effect of round-off errors. Choice (5.3.19) has been advocated in that it does satisfy the orthogonality condition $\bar{p}^{(i)T}y^{(i-1)} = 0$ even when the function is not quadratic. In fact there does seem to be evidence that when using alternative (ii), choice (5.3.19) is much superior to (5.3.20)—see Polak and Ribière (1969) for instance. However, alternative (ii) is not recommended, so the relevance of this evidence to reset algorithms is questionable. In fact, although choice (5.3.19) works better in some circumstances and choice (5.3.20) in others, I know of no systematic evidence which indicates how the choice should be resolved in a general purpose algorithm.

The advantage of the method of conjugate directions, as against those quasi-Newton methods which use first derivatives, is in the amount of computer storage ($\sim n$ as against $\sim n^2$ locations) and housekeeping time ($\sim n$ as against $\sim n^2$ operations/iteration) which is required. These factors are most important for very large problems, or if the function is very simple to calculate. However, my practical experience with the Fletcher–Reeves algorithm is that more iterations have usually been required for convergence as against the DFP method—a factor of 2 is typical. This has been ascribed to the fact that less information is stored in the Fletcher–Reeves method about the behaviour of the function. Nonetheless, the algorithm is extremely reliable and is well worth trying.

Another advantage of conjugate gradient methods is that convergence can be proved for general differentiable functions, in the sense either that $F \to -\infty$ or that $\|g^{(k')}\| \to 0$, where k' is an infinite subsequence of the iteration numbers. Furthermore, McCormick and Pearson (1969) show that for a wide class of *reset* conjugate direction methods like (5.3.21), superlinear convergence to a local minimum x^* can be proved in the sense that

$$\operatorname*{Lt}_{c\to\infty} \frac{\|x^{(nc+n+1)} - x^*\|}{\|x^{(nc+1)} - x^*\|} = 0, \tag{5.3.25}$$

providing that simple conditions on the search directions can be established. McCormick and Pearson have established these conditions for two of the algorithms and it seems likely that such a result can be proved for most reset algorithms. As for non-reset algorithms, the rate of convergence is certainly linear but whether it might be superlinear is currently the subject of some contention. However, in view of the unacceptable practical experience with these algorithms, the resolution of this point is more of interest to the theoreticians.

5.4 Minimization with first derivatives—other methods

A number of other gradient methods have been developed which are also based upon conjugacy properties and a brief survey of these will be given. The method suggested by Powell (1962) uses the parallel subspace property of Theorem 2 in a recursive fashion. For a quadratic function, the initial point $x^{(1)}$ can be considered as the minimum point in a plane whose normal is the gradient vector $g^{(1)}$. This plane is an $n-1$ dimensional subspace of E^n. If $x^{(2)}$ is taken as a point $x^{(1)}-\alpha g^{(1)}$ for $\alpha \neq 0$ (in fact Powell chooses the optimum point using a line search), then $x^{(2)}$ can be considered as an arbitrary point in a parallel $n-1$ dimensional subspace. If $x^{(2n-1)}$ is the optimum point in this subspace, then the line joining $x^{(1)}$ and $x^{(2n-1)}$ passes through the point which minimizes the function in E^n. Thus the problem of minimizing a function in E^n is reduced to that of minimizing a function in E^{n-1}. What happens therefore is that the first n steps consist of minimizations along gradient vectors in spaces of dimension $n, n-1, \ldots, 1$, and the final $n-1$ steps consist of minimizations along lines joining $(x^{(n-1)}, x^{(n+1)})$, $(x^{(n-2)}, x^{(n+2)})$, $\ldots$, $(x^{(1)}, x^{(2n-1)})$ in turn. Hence the method minimizes a quadratic function in $2n-1$ line searches. The method generalizes to solve non-quadratic functions merely by repeating the cycle of n iterations; it has disadvantages as against the method of conjugate gradients in that more line searches are required per cycle and the housekeeping is $O(n^3)$ operations per cycle as against $O(n^2)$ for conjugate gradients. No compensating advantages in empirical rates of convergence have been reported.

Another method is the Partan method (Shah *et al.*, 1964), and this method is closely related to the method of conjugate gradients. The iteration can be written as

$$x^{(2)} = x^{(1)} - \mu_1 g^{(1)} \tag{5.4.1}$$

and

$$\left.\begin{aligned} z^{(i)} &= x^{(i)} - \mu_i g^{(i)} \\ x^{(i+1)} &= z^{(i)} + \lambda_i (z^{(i)} - x^{(i-1)}) \end{aligned}\right\} i = 2, 3, \ldots, n, \tag{5.4.2}$$

where μ_i and λ_i are chosen by optimum line searches. The first iteration is therefore a steepest descent step after which the iteration takes the form of a search along the steepest descent vector, followed by a search along a line joining $z^{(i)}$ with $x^{(i-1)}$. To show the connection with the method of conjugate gradients, equations (5.4.2) can be written

$$x^{(i+1)} = x^{(i)} + \frac{1}{q_i}\{-g^{(i)} + e_{i-1}(x^{(i)} - x^{(i-1)})\}, \tag{5.4.3}$$

where $q_i = 1/(\mu_i(1+\lambda_i))$ and $e_{i-1} = \lambda_i/(\mu_i(1+\lambda_i))$. Clearly this is very similar to the statement of conjugate gradients implicit in (5.3.18). In fact Rutishauser (1959) shows that the method of conjugate gradients can be written in just this way if

$$q_i = g^{(i)T}Gg^{(i)}/g^{(i)T}g^{(i)} - e_{i-1} \tag{5.4.4}$$

and if

$$e_{i-1} = q_{i-1}g^{(i)T}g^{(i)}/g^{(i-1)T}g^{(i-1)}. \tag{5.4.5}$$

Now by virtue of the line search, μ_i is determined by $\mu_i = g^{(i)T}g^{(i)}/g^{(i)T}Gg^{(i)}$, whence (5.4.4) follows directly, and (5.4.5) can be obtained in a similar way by using the condition $g^{(i+1)T}(z^{(i)} - x^{(i-1)}) = 0$ which determines λ_i. Thus the points $x^{(i)}$ determined by Partan are the same as would be obtained using the method of conjugate gradients if the function were quadratic. Partan thus takes $2n-1$ line searches to minimize a quadratic function; it can be applied to general functions either by repeating the cycle of $2n-1$ iterations, or by continuing the two step iteration (5.4.2) indefinitely. The housekeeping and storage requirements are of the same order as for conjugate gradients; however to my knowledge no practical results have been produced to show that the method is consistently faster than the conjugate gradient method, and so the latter is preferable because it is more simple and minimizes a quadratic function in fewer searches.

Yet another method which utilizes the properties of conjugate gradients is due to Zoutendijk (see Zoutendijk, 1970, for instance). It is motivated by the fact that the direction $p^{(i)}$ minimizes p^Tp subject to the conditions (5.3.1) holding for all $j < i$, and subject to a normalization condition. In fact if the normalization condition is chosen to be $g^{(i)T}p^{(i)} = 1$, then the conditions (5.3.1) can be written

$$g^{(j)T}p^{(i)} = 1, \quad j = 1, 2, \ldots, i. \tag{5.4.6}$$

This led Zoutendijk to suggest a different but similar iteration in which $p^{(i)}$ is obtained either by minimizing $\|p\|_1$ or $\|p\|_\infty$ subject to the conditions (5.4.6). This requires the solution of a linear program at each iteration, in which advantage of certain special features can be taken to simplify the arithmetic. When applied to quadratic functions these methods still converge in n linear searches, though the sequence of points differs from that of the method of conjugate gradients. Zoutendijk suggests various ways in which the methods might be generalized to minimize general functions. However, to minimize a quadratic function requires $O(n^3)$ housekeeping operations, and in the absence of any contrary practical experience, the method of conjugate gradients is considered preferable.

Finally in this section, the relation of the method of conjugate gradients to some quasi-Newton methods is explored, and it is shown that other possible methods suggest themselves. If equation (5.3.4), that is $p^{(i)} = -Q^{(i)}g^{(i)}$ for generating the search directions is considered, then in addition to the properties of $Q^{(i)}$ outlined in (5.3.6)–(5.3.9) it will be noticed that the recurrence

$$Q^{(i+1)} = Q^{(i)} - \frac{(Q^{(i)}y^{(i)})(Q^{(i)}y^{(i)})^T}{y^{(i)T}Q^{(i)}y^{(i)}} \tag{5.4.7}$$

also holds. The use of this recurrence in conjunction with (5.3.4) yields another method which generates the same sequence of approximations $\{x^{(i)}\}$ as the conjugate gradient method, when applied to quadratic functions. One possible advantage which can be cited for such a method is that all the orthogonality conditions (5.3.1) are preserved, even when minimizing general functions. Pearson (1969) has compared this algorithm on difficult minimization problems which arise when using certain penalty functions, and concludes that it is preferable to the Fletcher–Reeves algorithm as regards the number of iterations required. Unfortunately the storage and housekeeping requirements are comparable to those of a quasi-Newton method, and DFP method in particular performs even better on these problems.

A suggestion which might lead to a better algorithm is made by Hestenes (1969). He points out that $Q^{(1)}$ need not be the unit matrix I, but may be any positive definite matrix; whence the use of (5.3.4) and (5.4.7) will still yield conjugate directions of search, although no longer those of section 5.3. Furthermore the fact that

$$\sum_{i=1}^{n} p^{(i)}p^{(i)T}/p^{(i)T}y^{(i)} = G^{-1} \tag{5.4.8}$$

for a quadratic function suggests that after a cycle based on (5.3.4) and (5.4.7), the $Q^{(1)}$ for the next cycle should be chosen by accumulating the left-hand side of (5.4.8). This introduction of a matrix which is related to G^{-1} makes the method very similar to the quasi-Newton class of methods. Clearly the method is related very closely to the DFP method, because if the individual terms on the left-hand side of (5.4.8) are added into $Q^{(i)}$ at each iteration, instead of being introduced *en bloc* after each cycle, then the recurrence becomes

$$Q^{(i+1)} = Q^{(i)} + p^{(i)}p^{(i)T}/p^{(i)T}y^{(i)} - Q^{(i)}y^{(i)}(Q^{(i)}y^{(i)})^T/y^{(i)T}Q^{(i)}y^{(i)},$$

which is precisely that used in the DFP method. No comparison of these methods has been made, but one noticeable advantage of the Hestenes approach is that line searches are not necessary in obtaining quadratic termination, so that the possibility exists of carrying out less function evaluations per iteration, and therefore of improving efficiency.

6. Quasi-Newton Methods

C. G. BROYDEN
Essex University

6.1 Basic philosophy

We begin by recalling the simplest form of Newton's method (this volume, Chapter 3) for solving the set of non-linear equations

$$f(x) = 0, \tag{6.1.1}$$

namely

$$x^{(k+1)} = x^{(k)} - (J^{(k)})^{-1} f^{(k)} \alpha^{(k)}, \tag{6.1.2}$$

where $x^{(k)}$ is the kth approximation to the solution, $f^{(k)}$ denotes $f(x^{(k)})$, $J^{(k)}$ denotes the Jacobian of $f(x)$ evaluated at $x^{(k)}$, and $\alpha^{(k)} = 1$ for all k. If the non-linear equations are obtained by setting the first partial derivatives of some other scalar function $F(x)$ equal to zero, then solving these equations is equivalent to finding a stationary value of $F(x)$. The Jacobian of $f(x)$ is in this case the Hessian matrix of $F(x)$ and is symmetric provided $F(x)$ and its first partial derivatives are continuous. If $F(x)$ is known to be convex then any stationary value of $F(x)$ is a minimum, and the Hessian is either positive definite or semidefinite. Thus provided the values of the first and second partial derivatives of $F(x)$ can be computed for any given x and the Jacobian never becomes singular, Newton's method may, in principle, be used to determine a minimum. Its advantages are that if it works at all then it works extremely well; convergence is rapid and in general is ultimately quadratic, and, if a sufficiently good initial estimate of the solution can be determined, it is probably the best available method. However, it is not without its disadvantages and we now consider three of these in particular.

Perhaps the most serious charge levelled against this form of the method is that it often fails to converge to a solution from a poor initial estimate. The successive iterates change in an apparently quite random fashion and not infrequently a value of $x^{(k)}$ is computed that leads to some kind of failure, e.g. the taking of a square root of a negative number, when evaluating $f(x)$. To overcome this problem we normally choose the parameter $\alpha^{(k)}$ so that in some sense $x^{(k+1)}$ may be said to be a better approximation to the solution of the problem than $x^{(k)}$. If for instance we are concerned with minimizing $F(x)$ we may choose $\alpha^{(k)}$ so that $F^{(k+1)} < F^{(k)}$, where $F^{(k)}$ denotes $F(x^{(k)})$. A choice of this nature, of course, requires for its implementation that some form of iteration or search procedure be carried out. It follows from equation (6.1.2) that this search is along a straight line in n-dimensional Euclidean space and hence this part of any optimization procedure is known as the "line search". If it is necessary to choose $\alpha^{(k)}$ to minimize $F(x)$ as a function of α then we say that an "exact line search" is required. However, the detailed choice of this scaling factor does not affect the basic philosophy behind the quasi-Newton methods and we will discuss it no further at this stage. We return to it later in connection with some particular algorithms.

The second disadvantage of Newton's method is the difficulty of evaluating $J(x)$ if $f(x)$ is a complicated function of x. In many industrial problems it is virtually impossible to obtain the elements of $J(x)$ as explicit expressions and even if it were possible it would, for some problems, be an extremely laborious and time-consuming operation. In these circumstances we are condemned to using some approximation B to the Jacobian, and equation (6.1.2) then becomes

$$x^{(k+1)} = x^{(k)} - B^{(k)-1} f^{(k)} \alpha^{(k)}. \tag{6.1.3}$$

One method of obtaining $B^{(k)}$ is by the use of forward differences, computing its ith column using the equation

$$B^{(k)} e_i = (f(x^{(k)} + h e_i) - f(x^{(k)}))/h, \tag{6.1.4}$$

where e_i is the ith column of the unit matrix of order n and h is a suitably small scalar. This, however, is an expensive use of machine time since in order to evaluate $J(x)$ in this somewhat simple-minded fashion it is necessary to compute the vector function $f(x)$, $n+1$ times. Since one value of $f(x)$ needs to be calculated anyway (in order to compute $x^{(k+1)}$ using equation (6.1.3)), this implies n additional evaluations of $f(x)$. A more sophisticated approach has been adopted by Brown and Conte (1968) which reduces the amount of additional labour by about half, but even in this case there is a very considerable penalty at each iteration. A further weakness of the forward difference approach is the essentially empirical choice of h. In the absence of rounding error h should be reduced to zero as the solution is approached (Samanski, 1967) but when rounding error is present, as it invariably is, h should in principle be chosen to minimize the sum of rounding and truncation errors. In practice it is usually chosen arbitrarily in the range 10^{-3} to 10^{-6}. Perhaps we should emphasize here though that these disadvantages are not sufficient to condemn methods of this type out of hand. It may well be possible that the total number of iterations using them is sufficiently small to make them competitive, and the evidence suggests that the more "difficult" the problem is, the more likely this reduction in the number of iterations becomes.

Another way of overcoming the difficulty of computing $J(x)$ is to evaluate an approximation to it numerically only at every m iterations, where m is some positive integer. The problem here lies in knowing how to choose m, and even if the best choice of m is made the method may still be inferior to methods of the quasi-Newton type. For these reasons this idea is hardly ever implemented today.

The third disadvantage of Newton's method is the necessity of solving a set of linear equations at each iteration. This is perhaps the least of the three disadvantages that we have recorded but it does take both extra time and extra programming, although these undesirable features are perhaps not so important now that we have very fast computers with large core stores.

We now look rather more closely at another way of overcoming the second disadvantage of Newton's method, and present the idea which provides the underlying motivation for the quasi-Newton methods. Let, then, B be some approximation to $J(x)$ and let us compute $x^{(k+1)}$ using equation (6.1.3). We now consider methods which enable us to obtain better approximations to $J(x)$ without any additional function evaluations. To expose the underlying idea behind these methods we consider $f(x)$ to be the linear function defined by

$$f(x) \equiv Ax - b, \tag{6.1.5}$$

where A is a constant matrix and b a constant vector. We note that in the context of function minimization this is equivalent to the assumption that the function to be minimized, $F(x)$, is given by

$$F(x) \equiv \tfrac{1}{2}x^T Ax - b^T x + c, \tag{6.1.6}$$

where A is a symmetric matrix (which is positive definite if $F(x)$ possesses a unique minimum), b a vector and c a scalar. Since in this case grad $F(x) \equiv Ax - b$, we can identify $f(x)$ with the gradient $F(x)$ and equation (6.1.1), which requires the gradient to be the null vector, becomes the condition for a stationary value of $F(x)$.

We return now to equation (6.1.5). It is clear that the Jacobian of $f(x)$ defined in this way is simply A, so that if $s^{(k)}$ and $y^{(k)}$ are defined by

$$s^{(k)} = x^{(k+1)} - x^{(k)}, \tag{6.1.7}$$

$$y^{(k)} = f^{(k+1)} - f^{(k)}, \tag{6.1.8}$$

then the Jacobian $J(x)$ satisfies the equation

$$J^{(k)} s^{(k)} = y^{(k)}. \tag{6.1.9}$$

Since $B^{(k)}$ is some approximation to $J^{(k)}$ we would like it also to satisfy equation (6.1.9), but since we cannot compute $f^{(k+1)}$ and hence $y^{(k)}$, until we have determined $B^{(k)}$ this is clearly impossible. We can, though, require that the *next* approximation to $J(x)$, namely $B^{(k+1)}$, satisfies

$$B^{(k+1)}s^{(k)} = y^{(k)}, \tag{6.1.10}$$

so that $B^{(k+1)}$ has at least one property of $J^{(k)}$. In the case where the functions $f(x)$ are non-linear so that $J(x)$ is not constant, and equation (6.1.9) ceases to be valid, the same sort of considerations apply. In equation (6.1.9) $J(x^{(k)})$ may be replaced by $\bar{J}^{(k)}$, where $\bar{J}^{(k)}$ is a matrix whose ith row is the ith row of $J(x^{(k)}+s^{(k)}\theta_i)$ for some θ_i satisfying $0 < \theta_i < 1$, so that $\bar{J}^{(k)}$ approximates $J(x^{(k)})$ to an accuracy which depends both on $\|s^{(k)}\|$ and the non-linearity of $f(x)$. The quasi-Newton equation in this case forces $B^{(k+1)}$ to assume one property of $\bar{J}^{(k)}$. Since for non-linear systems $J(x)$ (and hence $\bar{J}^{(k)}$) is not constant this ensures that the approximation B reflects changes in $J(x)$ and the hope is that this will assist in obtaining rapid convergence. Equation (6.1.10), the quasi-Newton equation, is the equation underlying all the methods discussed in section 6.3 (below). Since it does not define $B^{(k+1)}$ uniquely it applies to a class of methods where the properties of the individual methods vary with the particular choice of $B^{(k+1)}$. The methods that we discuss, however, have common properties other than that of satisfying the quasi-Newton equation, and we now turn our attention to these.

Clearly if $B^{(k)}$ is a reasonably good approximation to $J^{(k)}$ it is advantageous that $B^{(k+1)}$ should retain as far as possible the desirable properties of $B^{(k)}$. This suggests that $B^{(k+1)}$ is formed from $B^{(k)}$ by adding a correction term $C^{(k)}$ so that

$$B^{(k+1)} = B^{(k)} + C^{(k)} \tag{6.1.11}$$

and that some criteria (in addition to the requirement that the quasi-Newton equation be satisfied) be selected in order to more precisely determine $C^{(k)}$. The criteria that we consider here for specifying $C^{(k)}$, are

(1) $C^{(k)}$ to be a rank-1 matrix,

(2) $C^{(k)}$ to be a rank-2 matrix,

(3) $C^{(k)}$ to be a matrix of minimum norm.

Single-rank corrections have been considered by, among others, Broyden (1965), Davidon (1968), Murtagh and Sargent (1969) and Pearson (1969), double-rank corrections by Broyden (1970b), Davidon (1959), Fletcher and Powell (1963), Fletcher (1970b), Powell 1970b), and Pearson (1969), and minimum-norm corrections by Goldfarb (1970), and Greenstadt (1970). Since the majority of minimum-norm corrections are in fact either rank-1 or rank-2 corrections we will not consider them specifically from the minimum-norm point of view. We do remark, though, that the fact that the same algorithm may be derived by applying apparently unrelated criteria does inspire a certain amount of confidence in its use.

We have chosen here to explain the quasi-Newton methods by relating them to Newton's method, and this is representative of their historical development. But it is readily seen that *any* sequence of vectors $x^{(k)}$,no matter how it is generated, and the associated sequence of functions $f^{(k)}$ may be used, via equations (6.1.7) to (6.1.11), to construct approximations to the Jacobian of f. Moreover, this may be done in the case where f is an m-vector, x is an n-vector, and $m \neq n$, i.e. in the case where the Jacobian is rectangular. This case is of practical importance when solving the over-determined least-squares problem, where $m > n$, or in minimizing a function subject to equality constraints, where $m < n$. In both these cases matrix updates more commonly associated with Newton's method have been used to obtain estimates of the Jacobian, and examples of this usage are considered in section 6.3 (below).

This approach has been particularly successful for least squares problems. Here the basic iteration is the Gauss–Newton one (this volume, Chapter 3), where

$$x^{(k+1)} = x^{(k)} - (J^{(k)T}J^{(k)})^{-1}J^{(k)T}f^{(k)}. \tag{6.1.12}$$

The Jacobian has more rows than columns ($m > n$) but even so approximately satisfies equation (6.1.9). If the Jacobian is either difficult or expensive to evaluate then as before it may be replaced in equation (6.1.12) by an approximation $B^{(k)}$; and as before $B^{(k+1)}$ may be made to satisfy the quasi-Newton equation.

We have not yet suggested how the disadvantages of solving a set of linear equations at each step may be overcome and for some problems, particularly over-determined least-squares problems, no way has yet been found of avoiding this. In the case where the Jacobian matrix is square (and non-singular) and where Newton's iteration is used a technique has been devised to replace the solving of equations by a matrix-vector multiplication. This technique is based upon the observation that if a matrix is modified by adding a correction of rank r, then its inverse may also be modified by adding a correction of rank r. Thus instead of storing and modifying an approximation $B^{(k)}$ to the Jacobian it is sufficient to store and modify an approximation to the inverse Jacobian. If this is denoted by $H^{(k)}$, equation (6.1.3) becomes

$$x^{(k+1)} = x^{(k)} - H^{(k)}f^{(k)}\alpha^{(k)} \tag{6.1.13}$$

and we have the most common form of the iteration. Despite the successes achieved by algorithms of this kind over the last decade, development is still proceeding. If $H^{(k)}$ is symmetric and positive definite (as it is, in theory, for many optimization algorithms) so that $H^{(k)} = LL^T$ for some lower triangular matrix L, Gill and Murray (1971) have suggested storing and modifying the triangular factor L. This approach seems to give the most useful improvements if the gradient $f(x)$ is itself obtained using finite differences, and is dealt with more fully in this volume, Chapter 7.

6.2 General properties

In this section we consider both desirable and undesirable features of minimization algorithms and attempt to relate them to specific properties possessed by individual quasi-Newton methods. We begin by discussing briefly these general features that we would take into consideration when seeking an algorithm to solve a given problem.

Perhaps the most important requirement is that the algorithm should not converge to an incorrect solution. A particular example of this would be convergence to a saddle-point instead of a minimum, but since this defect is not peculiar to quasi-Newton methods we consider it no further. Another feature to be avoided at all costs is premature or false convergence, and here it is possible that certain properties of the quasi-Newton methods could give rise to failure. It sometimes occurs that the matrix H becomes singular or nearly so, and the resulting step-length, $\|s\|$, becomes in consequence negligible. If one terminated the iteration by testing $\|s\|$ alone then one could have false convergence, but this may be prevented by testing also $\|f\|$, the norm of the gradient. Of course the quantity of interest is the norm of the error but this can only be estimated. Let x^* be the solution of the system in question so that

$$f(x^*) = 0 \tag{6.2.1}$$

and make the assumption that in the neighbourhood of x^*, $f(x)$ is essentially linear, satisfying equation (6.1.5). Define the vector error e by

$$e = x - x^*. \tag{6.2.2}$$

Equations (6.1.5) and (6.2.1) yield

$$Ax^* - b = 0$$

so that, to a good approximation in the neighbourhood of x^*,

$$f(x) = Ae, \tag{6.2.3}$$

where $A = J(x^*)$. Assume that A is non-singular and define the matrix E by

$$E = B - A, \tag{6.2.4}$$

where B is the current approximation to A. If H is defined to be B^{-1} it follows from equations (6.2.3) and (6.2.4) that

$$e = (I - HE)^{-1} H, \tag{6.2.5}$$

so that, if $\|HE\| < 1$,

$$\|e\| \leqslant \frac{\|H\| \, \|f\|}{1 - \|HE\|}. \tag{6.2.6}$$

Thus if we suspect that $\|HE\| \ll 1$ we may, since $\|H\|$ and $\|f\|$ may be evaluated, estimate an upper bound for $\|e\|$. In order to use this device with some measure of confidence we would require that, once a good approximation to $J^{-1}(x^*)$ had been achieved, it would not be spoiled by subsequent iterations. We would hope at least that $\|E\|$ would not increase as the iteration proceeded and we would indeed prefer it to be reduced. Whether or not an algorithm possesses this property of matrix error norm reduction when it is used to minimize quadratic functions may be readily established theoretically. It is a property that influences both the convergence and stability properties of an algorithm and is thus a critical property to consider when attempting any assessment of merit.

Another property that we require of algorithms is that they should not fail catastrophically when updating the matrix H. Some algorithms have this property (in theory) at all stages of the process and some at no stage. Certain algorithms hope to avoid this failure since all divisors used in implementing the update are known in theory to be non-zero, and this type of failure hardly ever occurs. Even in those algorithms for which a zero division is theoretically possible, its occurrence is infrequent. A more usual cause of failure in the quasi-Newton algorithms is a tendency for the algorithm to get "stuck" at a certain stage of the process, when changes in the current approximation to the solution become negligibly small. This behaviour has been observed with many such algorithms (Bard, 1968; Pearson, 1969; Broyden, 1970b) and is nearly always associated with the occurrence of a singular H. To see why a singular value of H should cause this behaviour, we consider a general matrix updating formula which includes the most commonly used formulae as special cases.

For every algorithm discussed in the next section (with the exception of the Pearson–McCormick algorithms for which the following analysis does not apply) we may, since

$$s^{(k)} = -H^{(k)}f^{(k)}\alpha^{(k)},$$

write $H^{(k+1)}$ as

$$H^{(k+1)} = H^{(k)}M^{(k)}, \tag{6.2.7}$$

where $M^{(k)}$ is a matrix specific to a particular update. Thus, by induction,

$$H^{(k+r)} = H^{(k)}M^{(k)}M^{(k+1)} \dots M^{(k+r-1)}, \quad r \geqslant 1,$$

so that since

$$s^{(k+r)} = -H^{(k+r)}f^{(k+r)}\alpha^{(k+r)},$$

$$s^{(k+r)} = -H^{(k)}v, \quad r \geqslant 1, \tag{6.2.8}$$

where

$$v = M^{(k)}M^{(k+1)} \dots M^{(k+r-1)}f^{(k+r)}\alpha^{(k+r)}.$$

Suppose now that $H^{(k)}$ is singular, so that for some vector q, $q^T H^{(k)} = 0$. It follows immediately from equation (6.2.8) that $q^T s^{(k+r)} = 0$, $r \geqslant 1$, so that once a particular $H^{(k)}$ becomes singular all subsequent steps are orthogonal to some fixed vector and hence are restricted to lie in a subspace of E^n. Unless the solution also lies in this subspace (and in general it will not) it will be completely unattainable subsequent to the occurrence of a singular H. Although this or similar behaviour has been observed in all the algorithms considered here some algorithms are more prone to this failing than others. It has generally been overcome in practice by "resetting" H to be (usually) the unit matrix after, say, every $2n$ iterations (Pearson, 1969). Another possibility is the use of the normal update after taking a step $s^{(k)}$ other than that given by equation (6.1.13) (Powell, 1970b).

If then algorithms exist that avoid all the pitfalls described above one might then think in terms of obtaining more rapid convergence. Since Newton's method converges rapidly near the solution one might require an algorithm to resemble Newton's method as closely as possible, and this might lead to the requirement that the matrix error norms $\|E^{(k)}\|$, defined by equation (6.2.4), decrease in some way when minimizing a quadratic function. This property generalizes in the case of non-quadratic functions to the property of "bounded deterioration" (Dennis, 1971). If we define $E^{(k)}$ by

$$E^{(k)} = B^{(k)} - J(x^*),$$

where x^* is the solution to which the algorithm is converging, then for algorithms possessing the property of bounded deterioration some norm of $E^{(k)}$ increases after a number of steps by an amount not exceeding some constant times the sum of the step-lengths. Thus if $\|E^{(0)}\|$ is sufficiently small and $x^{(0)}$ is sufficiently close to x^* the amount of deterioration in the accuracy of the approximation to the Jacobian during the course of the iteration will be insufficient to prevent convergence. Thus if the property of bounded deterioration can be established for an algorithm when applied to a class of problems the construction of a local convergence proof is a formality.

Another property that might lead to rapid convergence overall is the property of minimizing a quadratic function in at most n steps. The desirability or otherwise of this property of "quadratic termination" and the analogous properties for solving general non-linear systems and constrained optimization problems (linear termination and quadratic/linear termination) has not been fully established, and a certain amount of uncertainty still remains. Fletcher (1970a) writes "These examples, and many others, suggest that the property of quadratic/linear termination is a desirable attribute to be aimed at when designing methods for non-linear programming", but we have, on the other hand, "The property of quadratic termination, whose relevance for general functions has always been questionable, . . ." (Fletcher, 1970b). The present author believes that the property of quadratic/linear termination is important provided that it is not achieved at the expense of stability, although he acknowledges that the evidence for this belief is stronger in the case of algorithms designed to solve general non-linear simultaneous equations than it is in the case of algorithms for function minimization.

Another feature that affects the overall speed of an algorithm is the amount of work required during each iteration. If $\alpha^{(k)}$ is taken to be unity in equations (6.1.3) and (6.1.13) then the amount of work is minimal. On the other hand, if it is necessary to minimize $F(x)$ then the computing cost of each iteration will be quite high since an inner iteration is needed at every step of the main iteration in order to compute $\alpha^{(k)}$. If this inner iteration has to be carried out at all it should clearly be carried out as efficiently as possible, and the precise mechanism by which this is achieved is beyond the scope of this chapter. However, a necessary feature for an efficient line search is the requirement that the vector $-H^{(k)}f^{(k)}$ always points in the downhill (or uphill) direction so that it is known before the search commences whether a positive (or negative) value of $\alpha^{(k)}$ will give the required minimum. This knowledge may then be used in initiating the search for the minimum and may well contribute in the subsequent choice of search strategy. We show now, therefore, that if $H^{(k)}$ is positive definite and $x^{(k+1)}$ is given by equation (6.1.13) then, for $\alpha^{(k)}$ sufficiently small and positive, we have $F^{(k+1)} < F^{(k)}$.

Let p be an arbitrary vector and α a scalar. Then, since

$$f(x) = \text{grad } F(x),$$

$$F(x^{(k)}+p\alpha) = F^{(k)}+\alpha p^T f^{(k)}+O(\alpha^2). \tag{6.2.9}$$

If we assume that $|\alpha|$ is sufficiently small so that terms in α^2 may be ignored (and we can always find such an α provided that f is continuous in a neighbourhood of $x^{(k)}$) we have, on substituting $-H^{(k)}f^{(k)}$ for p and appealing to equations (6.1.13) and (6.2.9),

$$F^{(k+1)}-F^{(k)} = -\alpha^{(k)}f^{(k)T}H^{(k)}f^{(k)}.$$

Thus if $H^{(k)}$ is positive definite a positive (if small) value of $\alpha^{(k)}$ will result in a reduction of $F(x)$ so that, provided $F(x)$ is continuous, a minimum of $F(x)$ will occur for a positive value of α. If we choose $\alpha^{(k)} = 1$ for all k there is no guarantee in general that the process defined by equation (6.1.13) and the relevant updating formulae will converge at all so that the additional work involved in minimizing $F(x)$ may be regarded as a premium to be paid in order to improve reliability. Recent tendencies, however (Fletcher,

1970b), have been to choose a value of $\alpha^{(k)}$ that merely *reduces* $F(x)$ despite the fact that this results in sacrificing the property of quadratic termination which relies, for many algorithms, upon exact minimization.

This last strategy illustrates the compromises forced upon the numerical analyst. No single algorithm embraces the properties of stability, quadratic termination and non-iterative determination of α. A choice therefore has to be made and the overall performance of the algorithms studied. In this way we hope to establish which properties give rise to stable, rapidly converging algorithms so that we can devise future algorithms specifically with the idea of possessing these particular properties.

6.3 Particular algorithms

In this section we consider particular algorithms for minimizing functions in the light of the discussion of the previous two sections. We shall be principally concerned with properties such as quadratic termination, bounded deterioration and stability and shall endeavour to relate these properties to the observed performance of the algorithms under discussion. We deliberately refrain from a detailed discussion of algorithms for solving non-linear simultaneous equations and constrained optimization problems. In general we are content to give such algorithms only a brief mention unless they have been used in the context of unconstrained optimization or unless they form the basis of more sophisticated minimization algorithms.

A. Single-rank algorithms

A.1. Broyden's algorithm (1965). This algorithm was intended for the solution of general sets of non-linear equations but it apparently performs remarkably well when used to minimize functions (Himmelblau, to appear). The reason for its inclusion in this survey though is because of its relationship to algorithms that have been developed specifically to solve minimization problems. The update of the approximate Jacobian $B^{(k)}$ is given by

$$B^{(k+1)} = B^{(k)} - (B^{(k)}s^{(k)} - y^{(k)}) \frac{s^{(k)T}}{s^{(k)T}s^{(k)}}, \tag{6.3.1}$$

so that, if B is square and non singular, it follows from the Sherman–Morrison formula (see appendix) that

$$H^{(k+1)} = H^{(k)} - (H^{(k)}y^{(k)} - s^{(k)}) \frac{s^{(k)T}H^{(k)}}{s^{(k)T}H^{(k)}y^{(k)}}. \tag{6.3.2}$$

It may be readily verified from equations (6.2.4) and (6.3.1) that

$$E^{(k+1)} = E^{(k)}\left(I - \frac{s^{(k)}s^{(k)T}}{s^{(k)T}s^{(k)}}\right), \tag{6.3.3}$$

so that, for quadratic functions, the matrix error $E^{(k)}$ decreases monotonically. The method is one of bounded deterioration and may be shown to be locally convergent for $\alpha^{(k)} = 1$ (Dennis, 1971). The algorithm is stable provided that $B^{(k)}$ approximates the Jacobian $J^{(k)}$ sufficiently well (Broyden, 1970a), but if the approximation is poor then the algorithm is unstable and its performance suffers in consequence. It is thus necessary to provide this algorithm with a starting procedure to ensure that the initial approximation is tolerably accurate. The algorithm does not enjoy the property of quadratic termination even when used for function minimization with exact line searches. There is thus no theoretical justification for carrying out exact line searches and $\alpha^{(k)}$ is usually chosen to reduce $\|f\|$ (Broyden, 1965), or is set to be equal to unity (Broyden, 1970a). The approximation H is not symmetric and neither is $H+H^T$ necessarily positive definite so, if line searches are used, this fact must be recognized by the line-search routine.

In practice, if the algorithm is in the region of local convergence with $\alpha^{(k)} = 1$ then it performs very well, convergence being usually super-linear. Broyden (1970a) showed that if f and e are given by equation (6.1.5) and (6.2.2), $\alpha^{(k)} = 1$ for all k and $\|E^{(0)}\| < 1$ then

$$\|e^{(r)}\| \leqslant (c/r^{\frac{1}{2}})^r \|e^{(0)}\| , \tag{6.3.4}$$

where

$$c = \frac{\|E^0\|_{\text{Euclidean}}}{1-\|E^{(0)}\|},$$

and $\|.\|$ denotes the L_2 norm.

A similar rate of convergence is observed for non-linear problems near the solution.

A.2. The secant algorithm (*Wolfe, 1959; Barnes, 1965*). This, the oldest of all the quasi-Newton methods (it was known in a rudimentary form to Gauss, 1809) is another method that is not restricted to the minimization of functions but may be used to solve a general set of non-linear equations. The updating equation is

$$B^{(k+1)} = B^{(k)} - (B^{(k)}s^{(k)} - y^{(k)}) \frac{q^{(k)T}B^{(k)}}{q^{(k)T}B^{(k)}s^{(k)}}, \tag{6.3.5}$$

where $q^{(k)}$ is given by

$$q^{(k)T}y^{(j)} = 0, \quad k-n+1 \leqslant j \leqslant k-1. \tag{6.3.6}$$

Hence, if B is square and non-singular,

$$H^{(k+1)} = H^{(k)} - (H^{(k)}y^{(k)} - s^{(k)}) \frac{q^{(k)T}}{q^{(k)T}y^{(k)}}.$$

Since $q^{(k)}$ is only defined by the above equations if at least $n-1$ steps have taken place a starting procedure is needed (e.g. Barnes, 1965).

The algorithm as described has the property of linear termination since it follows from equations (6.3.5) and (6.3.6) (after some manipulation) that

$$B^{(k+1)}s^{(j)} = y^{(j)}, \quad k-n+1 \leqslant j \leqslant k. \tag{6.3.7}$$

Thus $B^{(k+1)}$ is determined uniquely by the previous n steps so that for linear functions the approximation B becomes equal to the exact Jacobian after n steps. A further Newton step then gives the exact solution, and we have termination after $n+1$ steps.

The difficulty with the secant method lies in the probability, for non-linear problems, that n consecutive y's or n consecutive s's will become linearly dependent. This deprives it of the property of bounded deterioration and in practice makes it notoriously unstable. Attempts to reduce this instability are usually based on requiring equation (6.3.7) to be satisfied not for $k-n+1 \leqslant j \leqslant k$ but for n values of j chosen with stability in mind. Instead of replacing the "oldest" (s, y) pair by the "newest" (as equation (6.3.7) requires) the "newest" pair replaces a pair chosen so that the replacement impairs stability by the smallest amount. Thus the strict linear termination property is sacrificed to improve stability. The secant algorithm, modified or unmodified, may be used with $\alpha^{(k)} = 1$ or with $\alpha^{(k)}$ chosen to reduce $\|f\|$, but the method does not appear to be extensively used in either of these modes. As with the previous method its interest, in the present context, lies in the application of its basic philosophy to algorithms more specifically orientated towards optimization.

A.3. The McCormick–Pearson methods (*Pearson, 1969*). This class of methods is a generalization of two methods given by Pearson (1969), one of which he attributes to McCormick. It is a sub-class of the 3-parameter family of Huang (1970), for which

$$H^{(k+1)} = H^{(k)} + \rho^{(k)}s^{(k)}q_1^{(k)T} - H^{(k)}y^{(k)}q_2^{(k)T},$$

where

$$q_i^{(k)} = c_i s^{(k)} + k_i H^{(k)T} y^{(k)}, \quad i = 1, 2.$$

The scalar $\rho^{(k)}$ is arbitrary and the scalars c_i and k_i are also chosen arbitrarily but subject to the requirement that $q_i^{(k)T}y^{(k)} = 1$, $i = 1, 2$. It follows immediately that $H^{(k+1)}y^{(k)} = s^{(k)}\rho^{(k)}$ so that if $\rho^{(k)} = 1$ then $B^{(k)}$ must satisfy equation (6.1.10). Huang showed that if $x^{(k+1)}$ is given by

$$x^{(k+1)} = x^{(k)} - H^{(k)T}f^{(k)}\alpha^{(k)} \tag{6.3.8}$$

for all k and if $\alpha^{(k)}$ is always chosen to minimize $F(x)$ (exact line search) then all algorithms belonging to the class have the property of quadratic termination for any choice of the (essentially three) arbitrary parameters. Moreover, the steps $s^{(k)}$ satisfy the equation

$$s^{(k)T}As^{(j)} = 0, \quad j \neq k, \tag{6.3.9}$$

where A is the matrix defining the quadratic form given by equation (6.1.6), so that the methods generate conjugate directions and terminate, for quadratic functions, after at most n and not $n+1$ steps.

The update for the McCormick–Pearson method is given by

$$H^{(k+1)} = H^{(k)} - (H^{(k)}y^{(k)} - s^{(k)})\left[\frac{\gamma^{(k)}s^{(k)T}}{s^{(k)T}y^{(k)}} + \frac{(1-\gamma^{(k)})y^{(k)T}H^{(k)}}{y^{(k)T}H^{(k)}y^{(k)}}\right], \quad (6.3.10)$$

where $\gamma^{(k)}$ is an arbitrary parameter which is zero for Pearson's and unity for McCormick's algorithm. Pearson proved that the algorithms are stable when minimizing a quadratic function but he gave no proof of stability for general functions. It is likely that both McCormick's and Pearson's versions are of bounded deterioration and are locally convergent. The two versions were tested by Pearson (1969), and their performance was comparable with other methods.

A.4. The symmetric algorithm (*Davidon, 1968*). This algorithm, which has been investigated by Davidon (1968) and Murtagh and Sargent (1970), is the maverick not because it refuses to be categorized but because it fits into so many categories. It is a member of the McCormick–Pearson class, a (degenerate) member of the rank-2 single-parameter family (see below) and it has some of the features of the secant method.

It is the only single-rank method which preserves the symmetry of $B^{(k)}$, and from this and the fact that $B^{(k+1)}$ must satisfy equation (6.1.10) it is readily deduced that

$$B^{(k+1)} = B^{(k)} - \frac{u^{(k)}u^{(k)T}}{u^{(k)T}s^{(k)}} \quad (6.3.11)$$

and

$$H^{(k+1)} = H^{(k)} - \frac{v^{(k)}v^{(k)T}}{v^{(k)T}y^{(k)}}, \quad (6.3.12)$$

where

$$u^{(k)} = B^{(k)}s^{(k)} - y^{(k)}$$

and

$$v^{(k)} = H^{(k)}y^{(k)} - s^{(k)}.$$

The basic method is not of bounded deterioration and is notoriously unstable in its unmodified form. Davidon suggests various updating strategies and makes a choice after performing certain tests, and Murtagh and Sargent use a different reset strategy if a certain criterion is not satisfied. These devices enormously improve the reliability of the algorithm and good results have been obtained in both cases.

A possible explanation for the success of the symmetric algorithm may lie in the fact that exact line searches are not required in order to achieve quadratic termination Indeed, as was pointed out by Wolfe (1967), it is not even necessary to choose $x^{(k+1)}$ by equation (6.1.3) in order that in the quadratic case the update should in general give H to be the *exact* inverse Jacobian after n steps. This is in complete contrast with other algorithms of the McCormick–Pearson family (A.3, above) and the rank-2 single-parameter family (B.1, below), and is more reminiscent of the secant method (A.2, above). This enables considerable economies to be made when carrying out line searches, and these have contributed materially to the observed success of the algorithm.

A.5. Powell's method for sums of squares (*Powell, 1965*). This method is one in which the objective function $F(x)$ is given as a sum of squares, i.e.

$$F(x) = f(x)^T f(x), \tag{6.3.13}$$

and where it is assumed that $J(x)$, the Jacobian of $f(x)$, is not explicitly available. The method is essentially the Gauss-Newton method (Section 6.1 above, and Chapter 3 of this volume) but using a modified secant update (A.2, above) to obtain an approximation to $J(x)$. Instead of requiring the approximate Jacobian $B^{(k+1)}$ to satisfy (6.1.10), however, Powell requires that

$$B^{(k+1)}s^{(k)} = d^{(k)}\|s^{(k)}\|, \tag{6.3.14}$$

where $d^{(k)}$ is a more sophisticated approximation to the directional derivatives along $s^{(k)}$ at $x^{(k+1)}$ than $y^{(k)}/\|s^{(k)}\|$. Since $d^{(k)}$ is computed on the assumption that $F(x)$ is minimized along $s^{(k)}$ exact line searches are required. In practice the algorithm has performed successfully on a large number of problems, although it has been known to converge prematurely to a non-solution (Powell, 1970).

A.6. Powell's hybrid method (*Powell, 1970a, and this volume, Chapter 3*). This algorithm is intended for solving the non-linear simultaneous equations $f(x) = 0$ and it is not applicable to the general non-linear least squares problem. It is thus properly outside the scope of the review, but since it uses a "least-squares" approach we accord it a brief mention. If we define $F(x)$ by $f(x)^T f(x)$ and apply the method of steepest descent to $F(x)$ our direction of search is along the direction $-J(x)^T f(x)$, where $J(x)$ is the Jacobian of $f(x)$. If, on the other hand, we elect to try a Newton step to solve $f(x) = 0$ our direction of search is along $-J^{-1}(x)f(x)$. Powell therefore suggests using a linear combination of these two directions as a search direction and since the Jacobian is not assumed to be available he approximates both to $J(x)$ and $J^{-1}(x)$, and generally updates these approximations by equations (6.3.1) and (6.3.2). The final form of the algorithm embodies many checks and safeguards to ensure convergence and stability, the full details of which may be found in the original paper.

We now come to the double-rank algorithms. All of these are intended to be used only for function minimization so that no disadvantage accrues from the symmetry of the approximations. All of them, in fact, may be generated by symmetrizing the appropriate rank-1 algorithm.

B.1. Powell's symmetric algorithm (*Powell, 1970d*). This was the first algorithm to be obtained using a symmetrization technique, which Powell applied as follows. Let $B^{(k)}$ be symmetric and obtain $\bar{B}_1^{(k+1)}$ by using Broyden's formula

$$\bar{B}_1^{(k+1)} = B^{(k)} - (B^{(k)}s^{(k)} - y^{(k)}) \frac{s^{(k)T}}{s^{(k)T}s^{(k)}}. \tag{6.3.15}$$

Since $\bar{B}_1^{(k+1)}$ is not symmetric Powell defines $B_1^{(k+1)}$ by

$$B_1^{(k+1)} = \tfrac{1}{2}(\bar{B}_1^{(k+1)} + \bar{B}_1^{(k+1)T}). \tag{6.3.16}$$

However, $B_1^{(k+1)}$ will not satisfy

$$B_1^{(k+1)}s^{(k)} = y^{(k)}, \tag{6.3.17}$$

so that $B_1^{(k+1)}$ is corrected using equation (6.3.15) with $B^{(k)}$ replaced by $B_1^{(k+1)}$. The resulting matrix may then be symmetrized by an equation analogous to equation (6.3.16) to give $B_2^{(k+1)}$, and the whole process repeated indefinitely. The sequence of matrices $B_i^{(k+1)}$ converges as $i \to \infty$ to the limit $B^{(k+1)}$, where

$$B^{(k+1)} = B^{(k)} + \frac{1}{s^{(k)T}s^{(k)}} [(y^{(k)} - B^{(k)}s^{(k)})s^{(k)T} + s^{(k)}(y^{(k)} - B^{(k)}s^{(k)})^T] - \frac{s^{(k)}s^{(k)T}(y^{(k)} - B^{(k)}s^{(k)})s^{(k)T}}{(s^{(k)T}s^{(k)})^2} \tag{6.3.18}$$

and this gives a new updating formula for the approximation to the Jacobian that both satisfies equation (6.1.10) and is symmetric. In order to avoid the solution of equations $H^{(k)}$, the inverse of $B^{(k)}$, is stored and updated by a double-rank formula derived from equation (6.3.18).

A feature of this method is the behaviour of the matrix error when the algorithm is applied to the quadratic function defined by equation (6.1.6). If the $E^{(k)}$ is defined by equation (6.2.4) then for this algorithm

$$E^{(k+1)} = P^{(k)}E^{(k)}P^{(k)}, \tag{6.3.19}$$

where

$$P^{(k)} = I - \frac{s^{(k)}s^{(k)T}}{s^{(k)T}s^k}.$$

Thus for the single-rank update (A.1) the matrix error is post-multiplied by a projection matrix, whereas for the double-rank variation the error is both post- and pre-multiplied by the same projection matrix. It may be shown (Dennis, 1971) that the update is of bounded deterioration and is stable if $B^{(k)}$ approximates sufficiently closely to $J^{(k)}$. Algorithms based on using this update with equation (6.1.13) do not possess the property of quadratic termination for any choice of $\alpha^{(k)}$ so that there is no theoretical reason for using exact line searches. Indeed there is evidence (Dennis, 1971) that if $F(x)$ is minimized at each step the update behaves rather badly. If $\alpha^{(k)}$ is chosen to be unity for all k the performance of the update is comparable with that of Broyden (A.1), and the expectation that it would prove superior due to a greater reduction (in general) of the norm of $E^{(k)}$ at each stage has so far proved to be unfounded. In the algorithm proposed by Powell there are, as in the previous algorithm (A.6), many checks and safeguards incorporated to guarantee stability and convergence, but as the algorithm is comparatively recent there have as yet been few comparisons with other algorithms.

B.2. The single-parameter rank-2 family (*Broyden, 1967*). This family includes most of the better-known optimization algorithms, e.g. the Davidon–Fletcher–Powell (DFP) algorithm, as special cases. It is that subclass of Huang (1970) family where the correction to $H^{(k)}$ is symmetric and rank 2, and where $H^{(k+1)}$ is constrained to satisfy the equation

$$H^{(k+1)}y^{(k)} = s^{(k)}. \tag{6.3.20}$$

This is, since $H = B^{-1}$, a re-expression of equation (6.1.10). Since $H^{(k)}$ is symmetric for all k the equation for $x^{(k+1)}$ given by Pearson (1969) and Huang (1970), i.e. equation (6.3.8), reduces to equation (6.1.13). Provided that $x^{(k+1)}$ is always calculated using this equation, and that exact line searches are employed, the successive steps $s^{(j)}$ are conjugate for quadratic functions and we have quadratic termination for all algorithms included in the family. It has recently been shown (Broyden *et al.*, to appear) that certain members of the class enjoy the property of bounded deterioration and that if they are employed to minimize quadratic functions with $\alpha^{(k)} = 1$ for all k the vector errors satisfy equations similar to equation (6.3.4). Moreover, if these algorithms are applied with $\alpha^{(k)} = 1$ to more general functions, local convergence proofs may be derived. The case when exact line searches are used has been analysed extensively by Powell (1969, 1971) who proved that the DFP algorithm converges if the objective function $F(x)$ is convex, and converges superlinearly if $F(x)$ is uniformly convex. That these proofs extend to nearly all algorithms of the Huang class follows from a remarkable theorem proved by Huang (1970) for the case where $F(x)$ is quadratic, postulated by Huang and Levy (1970) for a general $F(x)$ and proved for general $F(x)$ by Dixon (1971). Dixon showed that nearly all algorithms in the Huang class that satisfy equation (6.3.20) will, given identical initial conditions and exact line searches, produce for any arbitrary function the identical sequence of $x^{(k)}$'s. Thus it is only necessary to establish convergence for one algorithm to have demonstrated convergence for the class as a whole.

The updating equation for the rank-2 family is

$$H^{(k+1)} = H^{(k)} - \frac{H^{(k)}y^{(k)}y^{(k)T}H^{(k)}}{y^{(k)T}H^{(k)}y^{(k)}} + \frac{s^{(k)}s^{(k)T}}{s^{(k)T}y^{(k)}} + \rho^{(k)}(H^{(k)}y^{(k)} - s^{(k)}\theta^{(k)})(H^{(k)}y^{(k)} - s^{(k)}\theta^{(k)})^T, \tag{6.3.21}$$

where

$$\theta^{(k)} = y^{(k)T}H^{(k)}y^{(k)}/s^{(k)T}y^{(k)}$$

and $\rho^{(k)}$ is arbitrary. It was shown by Broyden (1970b) that if $H^{(k)}$ is positive-definite then $H^{(k+1)}$ is also positive definite for $\rho^{(k)} \geqslant 0$ *provided that an exact line search has been carried out.* This result was extended by Shanno (1970) who proved that $H^{(k+1)}$ is positive definite for $(\rho^{(k)} > -\delta^{(k)}$, where $\delta^{(k)}$ is some positive number whose value depends upon $H^{(k)}$, $y^{(k)}$ and $s^{(k)}$. Since these results require only that $F(x)$ and its gradient are continuous they imply that algorithms of this family for which $\rho^{(k)} \geqslant 0$ should be stable for all reasonable problems, since all divisors occurring in equation (6.3.21) are theoretically positive. This prediction has in general been borne out in practice. The algorithm of Davidon (1959) as modified by Fletcher and Powell (1963), for which $\rho^{(k)} = 0$ for all k, has been used extensively and in general has shown acceptable stability properties. A more recent update (Goldfarb, 1970; Broyden, 1970c; Fletcher, 1970b; Shanno, 1970), is obtained when $\rho^{(k)}$ is given by

$$\rho^{(k)} = \frac{1}{y^{(k)T}H^{(k)}y^{(k)}}. \tag{6.3.22}$$

It follows that $\rho^{(k)}$ is positive if $H^{(k)}$ is positive definite, and this algorithm is also generally stable although like the DFP algorithm there have been instances where it has failed to converge, probably due to $H^{(k)}$ becoming singular. Various explanations have been offered for this departure from the theoretically predicted behaviour, and in particular Bard (1968) pointed out that poor scaling could cause $H^{(k)}$ to become singular. Recent work, however, by Abbott (1971) would appear to indicate that a more probable cause of loss of positive definiteness is failure to perform an exact line search. He showed that if $\rho^{(k)} \geqslant 0$ and $H^{(k)}$ is positive definite a necessary and sufficient condition for $H^{(k+1)}$ to be positive definite is that

$$s^{(k)T}y^{(k)} > 0, \tag{6.2.23}$$

and demonstrated that commonly-used line-search procedures and termination criteria could cause inequality (6.2.23) to be violated when solving certain problems involving penalty functions. The importance of an exact line-search is underlined by Dixon's theorem. Different members of the class have given in practice widely different results for the same problem starting with the same initial conditions, but a series of very careful experiments by Huang and Levy (1970) showed that these discrepancies disappear when sufficient care is taken in minimizing $F(x)$. We thus see that for members of the family (B.2) not only are exact line searches necessary for quadratic termination but they are also desirable for stability. For this and other reasons it is possible that algorithms similar to (B.1) above might be increasingly used for unconstrained minimization problems.

C. Other applications

We mention here very briefly some more of recent algorithms that use quasi-Newton techniques. The first is a method by Brown and Dennis (1971) for solving the non-linear least squares problem and seems particularly suited to the case where the minimum sum of squares is large. In this method components of the Hessian of $f(x)^T f(x)$ are approximated and are updated by using either (A.1) or (B.1). Since the performance is identical in both cases the latter is preferred since storage is reduced due to the symmetry of the matrices concerned.

A second algorithm (Dennis, to appear) is used for solving the unconstrained optimization problem. It was first obtained by symmetrization, but instead of symmetrizing Broyden's update given by equation (6.3.1) and inverting the resulting formula to give the update for H, Dennis first inverted equation (6.3.1) to give equation (6.3.2) and symmetrized this. Since the operations of symmetrization and inversion do not commute Dennis arrived at a new formula where the update is given by

$$H^{(k+1)} = H^{(k)} - \frac{1}{s^{(k)T}H^{(k)}y^{(k)}}\left[v^{(k)}s^{(k)T}H^{(k)} + H^{(k)}s^{(k)}v^{(k)T} - H^{(k)}s^{(k)}\rho^{(k)}s^{(k)T}H^{(k)}\right] \quad (6.3.24)$$

where

$$v^{(k)} = H^{(k)}y^{(k)} - s^{(k)}$$

and

$$\rho^{(k)} = \frac{v^{(k)T}y^{(k)}}{s^{(k)T}H^{(k)}y^{(k)}}.$$

Dennis found that the performance of this update with $\alpha^{(k)} = 1$ was comparable with (B.1), and if exact line searches were carried out it still performed quite well whereas (B.1) failed. As the experimental testing was severely limited, Dennis made no claims for the method other than that it merited further consideration.

Further applications of the quasi-Newton principle appear in connection with the constrained minimization problem (this volume, Chapter 3). In minimizing $F(x)$ subject to the m $(< n)$ equality constraints $c(x) = 0$ the

$m+n$ equations to be solved are

$$J^T(x)z = f(x) \tag{6.3.25a}$$

and

$$c(x) = 0, \tag{6.3.25b}$$

where $f(x)$ is the gradient of $F(x)$, $J(x)$ is the Jacobian matrix of $c(x)$ and z is the vector (or order m) of Lagrangian multipliers. The Jacobian of the system (6.3.25) is

$$J_s(x) = \left[\begin{array}{c|c} G & J^T(x) \\ \hline J(x) & 0 \end{array}\right], \tag{6.3.26}$$

where G is a linear combination of the Hessian matrices of $F(x)$ and each individual constraint function. There is no reason in principle why the equations (6.3.25) could not be solved by any quasi-Newton method but it is hoped that special purpose updates retaining the null partition of $J_s(x)$ might exhibit superior convergence properties. Let

$$\begin{bmatrix} K^{(k)} & M^{(k)T} \\ M^{(k)} & 0 \end{bmatrix}$$

approximate to $J_s^{(k)}(x)$. Equation (6.1.10) may then be written

$$\begin{bmatrix} K^{(k+1)} & M^{(k+1)T} \\ M^{(k+1)} & 0 \end{bmatrix} \begin{bmatrix} x^{(k+1)} - x^{(k)} \\ z^{(k+1)} - z^{(k)} \end{bmatrix} = \begin{bmatrix} h^{(k+1)} - h^{(k)} \\ c^{(k+1)} - c^{(k+1)} \end{bmatrix} \tag{6.3.27}$$

where

$$h(x) \equiv J^T(x)z - f(x),$$

and an updating strategy chosen to ensure that this equation is satisfied.

Both the methods discussed below assume that $J(x)$ is explicitly available, though each method uses this information differently. Kwakernaak and Strijbos (1970) set $M^{(k+1)}$ equal to $J^{(k+1)}$ and obtain, from equation (6.3.27),

$$K^{(k+1)}(x^{(k+1)} - x^{(k)}) = h^{(k+1)} - h^{(k)} - J^{(k+1)T}(z^{(k+1)} - z^{(k)}). \tag{6.3.28}$$

$(K^{(k+1)})^{-1}$ is then computed from $(K^{(k)})^{-1}$ using equation (6.3.28) and the secant update, and the inverse of the approximate system Jacobian is then obtained using the formula

$$\left[\begin{array}{c|c} K & J^T \\ \hline J & 0 \end{array}\right]^{-1} \equiv \left[\begin{array}{c|c} K^{-1} - K^{-1}J^TBJK^{-1} & K^{-1}J^TB \\ \hline BJK^{-1} & -B \end{array}\right] \tag{6.3.29}$$

where

$$B = (JK^{-1}J^T)^{-1}.$$

Broyden and Hart (1970), use an approximation $M^{(k)}$ to $J^{(k)}$ when computing the inverse system Jacobian. Expanding equation (6.3.27) gives

$$M^{(k+1)}(x^{(k+1)} - x^{(k)}) = c^{(k+1)} - c^{(k)} \tag{6.3.30}$$

and

$$K^{(k+1)}(x^{(k+1)}-x^{(k)}) = h^{(k+1)}-h^{(k)}-M^{(k+1)T}(z^{(k+1)}-z^{(k)}) \quad (6.3.31)$$

and many combinations of update may now be used to compute $M^{(k+1)}$ and $K^{(k+1)}$. A typical, and quite effective one, is to obtain $M^{(k+1)}$ from equation (6.3.30) using the Broyden (A.1) update and to then compute $K^{(k+1)}$ using equation (6.3.31) and the Powell (B.1) update. Algebraic manipulation then yields a rank-2 updating formula for the approximate inverse system Jacobian.

This concludes our survey of quasi-Newton methods. We have attempted to assess the merits of the basic updates, used either in a "Newton" ($\alpha^{(k)} = 1$) mode or in a minimization mode, and to isolate those properties of the updates that contribute to its observed performance. We have not attempted to perform a consumer analysis upon the subroutines or procedures in which the updates are used since we are concerned primarily with the update itself. We realize that the reputation of an update may well be due as much to an artful choice of checks, safeguards and program constants with which it is surrounded as to the properties inherent in the update itself, and are only too conscious that a good update may be enhanced, and a poor one disguised, by such devices. These, however, we regard as generally beyond the scope o this survey and refer the reader to the original papers for fuller and more detailed descriptions.

7. Failure, the Causes and Cures

W. Murray
National Physical Laboratory

7.1 Introduction

There are many reasons why a user may fail to find the minimum of a function when using a minimization algorithm. It has been said that there are two kinds of failure. The first is the miserable failure which is discovered when an exasperated computer finally prints out a message of defeat. The second is the disastrous failure when the computer and trusting user mistakenly think they have found the answer. This chapter is concerned with some of the failures that may occur and how, if possible, to avoid them. An important point to remember is that errors can only be detected if suitable checks are made. Not all failures are mentioned but hopefully after reading this chapter the user will become more sceptical of computer results and published algorithms.

7.2 Programming Error

It is easy to realize that a user is unlikely to find the required answer by solving the wrong problem. The first thing that must be checked is whether the procedure to evaluate $F(x)$ does indeed give the correct value. This can often be done by knowing the value of the function for a particular value of x. Care should be taken however to ensure that the value of x at which the function is evaluated fully tests the program. An actual error made in evaluating a function which was not revealed by this test is illustrated below. In this example part of the evaluation of $F(x)$ required the evaluation of a scalar α where

$$\alpha = \sum_{i=1}^{n} T_1(x_i)$$

and $T_1(x_i)$ is the shifted Chebyshev polynomial of order 1 in the range $0 \leqslant x_i \leqslant 1$.

The initial value of x had the elements $x_i = i/(n+1)$. In the evaluation of $F(x)$, α was multiplied by a scalar β and an error had been made in the definition of β. This error was not revealed by evaluating $F(x)$ at our initial choice of x since at this value $\alpha = 0$.

It may be worthwhile at this stage to pay some attention to the accuracy to which the function is evaluated. Often an engineer only requiring, say, 4 figures in hisr esults, will write a procedure that evaluates the function to this accuracy. If an algorithm is then used which attempts to estimate derivatives by finite differences it is possible that there will be no figures of agreement between the actual gradient and the estimate obtained.

It is rare for the function to be programmed incorrectly, since function values are often known for non-trivial values of x. A much more common occurrence is for derivatives to be programmed incorrectly. If, as is often the case, no values are known for the first derivatives at a particular value of x then a means of checking the gradient at a cost of only one additional function evaluation is as follows. Evaluate

$$F(x^{(0)}+hp),$$

where $\|p\| = 1$ and h is some small scalar which is machine dependent, say

$$h = 2^{-t/2},$$

on a machine with a t bit wordlength. Then

$$\frac{F(x^{(0)}+hp)-F(x^{(0)})}{h} \doteq p^T g^{(0)}. \tag{7.2.1}$$

The vector p can be generated at random or chosen to be some specific vector such as

$$p = -g^{(0)}/\|g^{(0)}\|.$$

If (7.2.1) is not true then either there is a programming error in evaluating $g^{(0)}$ or a high level of rounding error in evaluating $F(x)$. This latter error can be checked by evaluating $F(x^{(0)}-hp)$ then checking to see if

$$F(x^{(0)}+hp)-F(x^{(0)}) \doteq \tfrac{1}{2}(F(x^{(0)}+hp)-F(x^{(0)}-hp)). \tag{7.2.2}$$

If (7.2.2) is true it is almost certain that there is a programming error in determining g. When the right-hand sides of (7.2.2) and (7.2.1) are approximately the same then g is probably programmed correctly. If this is not the case the whole test should be repeated with a larger value of h. Alternatively, steps could be taken to improve the accuracy to which $F(x)$ is evaluated.

Second derivatives of $F(x)$ can be checked in a similar manner once it is known that the gradient is programmed correctly. In this case the following relationship should hold

$$hp^T G^{(0)} p \doteq p^T(g^{(h)}-g^{(0)}), \tag{7.2.3}$$

where $g^{(h)}$ is the gradient of $F(x)$ at $x^{(0)}+hp$.

If (7.2.3) is not true then the test should be repeated with a larger value of h. Let h_1 and h_2 be the two successive values of h used. If (7.2.3) does not hold for either value of h and

$$h_2 p^T(g(h_1)-g^{(0)}) \doteq h_1 p^T(g(h_2)-g^{(0)}) \tag{7.2.4}$$

then it is likely that there is an error in the procedure that calculates $G^{(k)}$. If (7.2.4) does not hold then either the test should be repeated with a larger value of h or attention should be given to the accuracy to which g is calculated.

7.3 Rounding error

Unfortunately computers do not perform algebraic operations exactly, a point that is often not appreciated by authors and users of algorithms alike. It pays, therefore, to give some attention to the details of how an expression is calculated. A simple example to illustrate this point is given by

$$b = (1-(1-a)^{\frac{1}{2}})/a. \tag{7.3.1}$$

If $a = 0$ then b is indeterminate but almost certainly the value that is required is $b = \frac{1}{2}$ since

$$\lim_{a\to 0} b = \tfrac{1}{2}.$$

Any quantity being indeterminate at a point often indicates that it is being inaccurately evaluated in the neighbourhood of this point. There are two ways of rearranging the expression (7.3.1) to overcome this difficulty. The first is to express b as a series in a, that is

$$b = \tfrac{1}{2}+a/8+a^2/16+5a^3/128+7a^4/256+\dots$$

and to use this series if $|a| < 0{\cdot}01$.

The second is simply to write (7.3.1) equivalently as

$$b = 1/(1+(1-a)^{\frac{1}{2}}).$$

The first technique, although inferior to the second in this instance, can be applied in a wide variety of circumstances where error occurs due to cancellation between similar terms, both of which can be expanded in a series.

Another example is the expression (1.6.1) given for the minimum of a quadratic fitted through three points, namely

$$\hat{x} = \tfrac{1}{2}\left\{\frac{(x_2^2-x_3^2)F_1+(x_3^2-x_1^2)F_2+(x_1^2-x_2^2)F_3}{(x_2-x_3)F_1+(x_3-x_1)F_2+(x_1-x_2)F_3}\right\}. \tag{7.3.2}$$

It is easy to see that if $x_1 \doteq x_2 \doteq x_3$ then cancellation error will occur in (7.3.2). This is not however the most common source of error that can result for result from using (7.3.2). Suppose

$$F_1 \doteq F_2 \doteq F_3 \tag{7.3.3}$$

then it can be seen by rearranging (7.3.2) into the form

$$\hat{x} = \tfrac{1}{2}\left\{\frac{x_3^2(F_2-F_1)+x_1^2(F_3-F_2)+x_2^2(F_1-F_3)}{x_3(F_2-F_1)+x_1(F_3-F_2)+x_2(F_1-F_3)}\right\} \tag{7.3.4}$$

that this can also result in cancellation error.

Unfortunately (7.3.3) can be true (see Fig. (7.3.1)) even if the minimum lies in the interval (x_1, x_3) and neither x_1, x_2 or x_3 is close to the minimum. Consequently when interpolating, the use of (7.3.2) may lead to a predicted minimum that does not lie in the interval (x_1, x_3). The use of this formula

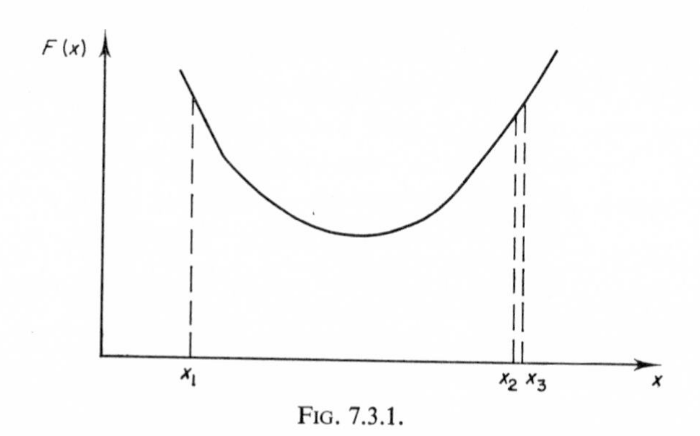

FIG. 7.3.1.

is therefore questionable when the function values are similar. This will be the case when $F(x)$ is slowly varying along the direction of search and must be ultimately true for every function if an accurate linear search is required. There is little that can be done to avoid this error. It emphasizes the need to include the checks suggested in Chapter 1. The difficulty will obviously be aggravated if apart from the error in calculating $\hat{x}$ there is a significant error in calculating $F(x)$.

7.4 Transformation of variables

In Chapter 3, Powell has described how some constrained minimization problems may be converted to an unconstrained problem by transforming the variables. Considerable care needs to be exercised to ensure that the transformation chosen does not create a difficult problem in the new variables. The following difficulties may be introduced by a transformation of variables:

(i) The function may be infinite at certain values of the new variables.
(ii) The degree of non-linearity of the function may be increased.
(iii) The Hessian matrix may become singular or ill-conditioned.
(iv) The function may become periodic in the new variables.

The last difficulty invariably occurs when trigonometrical transformations are used. Suppose the transformed variables are y with

$$F(y+j\alpha e_i) = F(y), \quad j = \pm 1, \pm 2, \ldots .$$

If in a minimization algorithm the predicted change to y_i is p_i, then p_i should be altered by subtracting or adding a multiple of α until

$$|p_i| < \alpha.$$

It is not easy to formulate rules that will always avoid (i)–(iii) but it has been our experience that trigonometrical and exponential transformations do create more difficulties than alternative transformations, especially for large n. Consider, for instance, two possible transformations that convert the problem

$$\text{minimize } \{F(x)\},$$

$$\text{subject to } \sum_{i=1}^{n} x_i^2 = 1,$$

into an unconstrained problem in the new variables y.

Transformation 1,

$$x_1 = \sin y_1 \ldots \sin y_{n-1},$$

$$x_i = \cos y_{i-1} \sin y_i \ldots \sin y_{n-1}, \quad i = 2, \ldots, n.$$

Transformation 2,

$$x_i = y_i/a, \qquad i = 1, \ldots, n-1,$$

$$x_n = 1/a,$$

where

$$a = \left(1 + \sum_{i=1}^{n-1} y_i^2\right)^{\frac{1}{2}}.$$

We have found the second transformation to be considerably superior to the first especially for large n. It is easy to see that the first transformation poses some difficulties when $y_{n-1} = 0$ since $F(y)$ is then invariant to changes in the other variables. Indeed, if any y_i, $i > 1$, is small or zero then $F(y)$ is almost invariant with respect to changes in at least one other variable.

7.5 Numerical stability

All the methods described in this volume are intended to be implemented on a computer. The arithmetic operations on a computer are not carried out exactly. They do, however, produce a result which corresponds to exact arithmetic with slightly perturbed operands. Thus the result of multiplying two scalars a and b on a computer is as if the operands had been $a+\varepsilon_1$ and $b+\varepsilon_2$ and the operation performed exactly. The scalars ε_1 and ε_2 are related to the wordlength of the computer employed. It is obviously important that the properties of a method are not dependent on exact arithmetic being performed. Alternatively it is important that the implementation of a method is such that it preserves the properties on which the method is based regardless of perturbations in the operands.

Since various authors interpret the term "numerical stability" in different ways it is perhaps useful to state what we understand by this term. Suppose an algorithm performs certain operations in order to determine the value of x, say $\hat{x}$, at which to carry out the next function evaluation. An algorithm is said to be numerically unstable if the process of these operations is such that it is possible for the value of x determined, say $\bar{x}$, to differ radically from $\hat{x}$. The numerical instability may either be inherent in the method because there is no guaranteed way of determining an accurate estimate to $\hat{x}$ or it could have been introduced by poor implementation. The fact that an algorithm is numerically unstable does not imply it will fail, although the introduction of random errors is unlikely to increase the probability of convergence. It is reasonable to suppose that the efficiency of an algorithm will be impaired if the numerical instability is such that it abrogates important properties of the method.

It is of some interest to examine the comparative merits of the algorithms described in this volume with regard to their numerical stability. The direct search algorithms are by their very nature numerically stable. Clearly an algorithm that does not require the function to be accurately computed is unlikely to depend on exact arithmetic utilizing these function values. The second derivative methods that are recommended in Chapter 4 are also very stable. This is due in part to the fact that the direction of search at each iteration is determined from information computed within that iteration.

This leaves only the conjugate direction and quasi-Newton methods. The recursive nature of these procedures introduces the possibility that an error made in one iteration may be reflected in all subsequent iterations. This could result in apparent inconsequential errors made in each iteration having a serious cumulative effect.

The conjugate gradient methods described in Chapter 5 are largely redeemed by the simplicity of the computation involved (no matrix vector multiplications) and the periodic resetting procedure. The numerical aspects of the conjugate direction method of Powell (1964) have been the subject of study by a number of workers, see for instance Brent (1971) and Rhead (1971). Difficulties arise in this method because the set of conjugate directions become nearly linearly dependent. Brent's solution is periodically to reset the conjugate directions to be a set of orthonormal directions based on the rejected conjugate directions. This preserves the property that the search directions span E^n and results in a more efficient algorithm than one using an arbitrary resetting procedure. Some information is still lost if the conjugate directions are almost linearly dependent. This is easily seen by considering the case when the conjugate directions are exactly linearly dependent. The loss of information would not be important if it were rare for the conjugate directions to degenerate, unfortunately when n is moderate or large it happens frequently even on functions whose Hessian matrix is well conditioned.

The only methods still to be considered are the quasi-Newton methods and these are dealt with in some detail in the next two sections.

7.6 Quasi-Newton algorithms

As with the conjugate direction procedures care has to be taken when implementing a quasi-Newton method. This can easily be seen by considering the consequences of $H^{(k)}$ becoming singular in the DFP method. The kth iteration is as follows (see Chapter 6),

$$\left.\begin{aligned} p^{(k)} &= -H^{(k)}g^{(k)}, \\ x^{(k+1)} &= x^{(k)} + \alpha^{(k)}p^{(k)}, \\ \text{and} \qquad H^{(k+1)} &= H^{(k)} - \frac{H^{(k)}y^{(k)}y^{(k)T}H^{(k)}}{y^{(k)T}H^{(k)}y^{(k)}} + \frac{\alpha^{(k)}p^{(k)}p^{(k)T}}{p^{(k)T}y^{(k)}} \end{aligned}\right\} \tag{7.6.1}$$

Clearly $p^{(k)}$ lies in the range of $H^{(k)}$ and this implies that the range of $H^{(k+1)}$ is a subspace of the range of $H^{(k)}$. It follows that the vector $p^{(k+1)}$ must also lie in the range of $H^{(k)}$, consequently if $H^{(k)}$ is singular the set of all subsequent search directions fails to span E^n. If $H^{(k)}$ had been calculated using exact arithmetic and $H^{(1)}$ was positive definite then $H^{(k)}$ could not be singular. Unfortunately these conditions are not met when implementing the DFP method on a computer with a finite word length.

The relative error made (Wilkinson, 1963; Gill, Murray and Pitfield, 1972) when multiplying a vector by a matrix may be large if the condition number of the matrix is large. Consequently, even if $H^{(k)}$ has been computed using exact arithmetic there may be a large relative error in computing $H^{(k)}y^{(k)}$. The error incurred in this computation will be amplified by the error made in computing $y^{(k)}$ which can be large since it is formed by differencing what are often two similar vectors. It follows that the computed scalar $y^{(k)T}H^{(k)}y^{(k)}$ may be negative or zero even when $H^{(k)}$ is positive definite.

The consequences of rounding errors made by computing $H^{(k+1)}$ and $p^{(k)}$ using (7.6.1) are summarized below where $\bar{H}^{(k+1)}$ denotes the computed $H^{(k+1)}$.

(i) The matrix $\bar{H}^{(k+1)}$ may not be positive definite.

(ii) The elements of $\bar{H}^{(k+1)}$ may differ in all figures from those of $H^{(k+1)}$.

(iii) The direction of search may not be a descent direction even if $\bar{H}^{(k)}$ is positive definite.

(iv) The computation of $\bar{H}^{(k+1)}$ may cause overflow.

(v) If $H^{(k)}$ is nearly singular subsequent approximations to G have a tendency to remain nearly singular.

Rounding errors, although unavoidable, can be minimized and it is possible to prevent their effect from being catastrophic. There have been a number of suggestions to circumvent some of the difficulties encountered with quasi-Newton methods. Pearson (1969) and Bard (1968) have proposed that the Hessian approximation should be periodically reset to some prescribed matrix. The disadvantage of the resetting schemes they suggest is that good information is often discarded and this can adversely affect the rate of convergence. Another possibility is periodically to determine the eigensystem of $H^{(k)}$ and form a new approximation by altering those eigenvalues that are very small or negative. This is an expensive procedure in terms of computer operations and storage.

All these schemes suffer from the disadvantage that it is not known when it is beneficial to reset $H^{(k)}$. An alternative implementation to (7.6.1) when $H^{(k)}$ is symmetric has been suggested by Gill and Murray (1971).

In this implementation the direction of search $p^{(k)}$ is found by solving the equations

$$B^{(k)}p^{(k)} = -g^{(k)}. \tag{7.6.2}$$

The matrix $B^{(k)}$ is recurred in the factorized form as

$$B^{(k)} = L^{(k)}D^{(k)}L^{(k)T}, \tag{7.6.3}$$

where $L^{(k)}$ is a lower-triangular matrix with unit diagonal elements
and $D^{(k)}$ is a diagonal matrix.
Equation (7.6.2) can be rewritten as

$$L^{(k)}D^{(k)}L^{(k)T}p^{(k)} = -g^{(k)},$$

from which $p^{(k)}$ can be found by solving successively

$$L^{(k)}v = -g^{(k)}$$

and

$$L^{(k)T}p^{(k)} = D^{(k)-1}v,$$

using forward and backward substitution. The motivation for the work of Gill and Murray is demonstrated in the following example which illustrates the difference in the effects of rounding error resulting from calculating a scalar α by two different methods.
Define α to be

$$\alpha = 1/v^TB^{-1}v,$$

where B is an $n \times n$ positive-definite symmetric matrix
and v is an $n \times 1$ vector.
It follows from the properties of positive-definite matrices that

$$\alpha > 0, \quad \text{if } v \neq 0.$$

Suppose we calculate an approximation, say $\bar{\alpha}$, to α using a t bit word-length computer in the following way. Let $fl(.)$ denote the result of floating point computation

$$w = fl(Hv),$$
$$\bar{\alpha} = fl(1/w^T v).$$

where $H = B^{-1}$.

Normally the elements of H and v will not be expressible in a t bit word-length number so that usually it will be necessary to use approximations to these quantities. Even ignoring this source of error $\bar{\alpha}$ no longer has the property,

$$\bar{\alpha} > 0, \quad v \neq 0.$$

If the triangular factorization LDL^T of B is known then $\bar{\alpha}$ can be computed by solving

$$Lu = v,$$

and setting

$$\bar{\alpha} = fl\left(1 \Big/ \sum_{i=1}^{n} d_i^{-1} u_i^2\right),$$

where d_i is the ith diagonal element of D.

The scalar $\bar{\alpha}$ approximates α and also retains the property that $\bar{\alpha} > 0$ if $v \neq 0$. This property is retained even if the factorization is not known exactly provided the approximations to d_i are positive. It is interesting to note that the second method of computing $\bar{\alpha}$ requires only $\frac{1}{2}n^2 + O(n)$ operations whilst the first requires $n^2 + O(n)$.

It is a property of the factorization (7.6.3) that if the diagonal elements of $D^{(k)}$ are positive $B^{(k)}$ is positive definite. It has been shown in Chapter 6 that $B^{(k+1)}$ is obtained from $B^{(k)}$ by the addition of a rank-2 matrix, that is

$$B^{(k+1)} = B^{(k)} + \pi_1 zz^T + \pi_2 ww^T, \tag{7.6.4}$$

where z and w are $n \times 1$ vectors and π_1, π_2 are scalars.

Gill and Murray show how the triangular factors are modified when a matrix is altered by a rank-1 matrix. Repeated application of their algorithm gives $L^{(k+1)}$ and $D^{(k+1)}$ from $L^{(k)}$ and $D^{(k)}$. A feature of their algorithm is that $D^{(k+1)}$ is guaranteed to have positive diagonal elements. It is also possible to ensure that $B^{(k+1)}$ is sufficiently positive definite to guarantee that $p^{(k+1)}$ is a descent direction even in the presence of rounding error. A further advantage of their implementation is that the vectors z and w can be obtained for many quasi-Newton methods without a matrix-vector product. If $B^{(k)}$ is theoretically guaranteed to be positive-definite then the alterations introduced to preserve the properties of the method are only at a level comparable to the rounding errors incurred. All the consequences (i)–(v) associated with quasi-Newton methods have been removed by this implementation.

For those methods for which $B^{(k)}$ is not theoretically guaranteed to be positive-definite it has been found (see Murtagh and Sargent, 1970) that a superior strategy is to alter the modification rule for those iterations for which $B^{(k)}$ would have become indefinite. This is a similar strategy to that recommended for the second derivative methods when the Hessian matrix was indefinite. The modification rule can be altered to maintain $B^{(k)}$ positive-definite simply by altering the scalars π_1 and π_2.

7.7 Quasi-Newton methods without derivatives

When analytical derivatives are not available we can, by replacing the gradient vector by a finite-difference approximation, still use a quasi-Newton method. The truncation error introduced by this approximation nullifies some of the theoretical properties of the quasi-Newton method. Since the gradient is zero at the solution the truncation error may be large relative to the gradient. Clearly care must be exercised in performing the finite-difference approximation. Two possible choices of finite-difference formulae for the approximate derivative $\bar{g}_i^{(k)}$ are

$$h_i^{(k)}\bar{g}_i^{(k)} = F(x^{(k)}+h_i^{(k)}e_i)-F(x^{(k)}) \tag{7.7.1}$$

and

$$2h_i^{(k)}\bar{g}_i^{(k)} = F(x^{(k)}+h_i^{(k)}e_i)-F(x^{(k)}-h_i^{(k)}e_i), \tag{7.7.2}$$

where e_i is the ith column of the identity matrix and $h_i^{(k)}$ is the ith element of the $n\times 1$ vector $h^{(k)}$. The leading terms of the truncation errors in $\bar{g}_i^{(k)}$ are multiples of $h_i^{(k)}$ and $(h_i^{(k)})^2$ respectively. If a quasi-Newton method has quadratic termination this property is preserved when $g^{(k)}$ is replaced by $\bar{g}^{(k)}$ defined by (7.7.2) but is not preserved if (7.7.1) is used. The advantage of using (7.7.1) is that it requires n less function evaluations to approximate $g^{(k)}$ compared with (7.7.2).

An algorithm employing difference approximations to the derivatives in conjunction with the DFP algorithm has been given by Stewart (1967). The basis of Stewart's paper is that the vector $h^{(k)}$ in (7.7.1) is chosen to balance truncation error against cancellation error. In general this implies $h^{(k)} \neq h^{(k+1)}$. If the total error in the approximation is too large Stewart switches to estimating $g^{(k)}$ by (7.7.2). Should the procedure fail to yield a descent direction of search the matrix $H^{(k)}$ is reset to I, the identity matrix.

Stewart recurs the diagonal elements of $B^{(k)}$ using the DFP formula. These elements are used as approximations to the diagonal elements of $G^{(k)}$ in order to estimate the truncation error. Unfortunately, although a matrix $B^{(k)}$ has certain properties similar to those of $G^{(k)}$, this does not in itself imply that

$$B_{i,i}^{(k)} \doteq G_{i,i}^{(k)}. \tag{7.7.3}$$

In Stewart's method, if (7.7.3) is not valid then a poor value of $h^{(k)}$ may be obtained. Murray (1972) has shown that the estimates to $G_{i,i}^{(k)}$ obtained from using different quasi-Newton formulae differ radically even when these methods generate the same sequence of $x^{(k)}$. In any event the elements of $B^{(k)}$ are unlikely to be estimates of $G^{(k)}$ if less than n iterations have been performed since the last reset. By using the DFP approximations on the diagonal elements of $G^{(k)}$ Stewart limits these estimates to be positive. A different recurrence approximation that did not inhibit the estimates to be positive would be more likely to give better results. It is certainly not necessary or desirable to reset the diagonal elements of $B^{(k)}$ just because $H^{(k)}$ is reset.

When choosing $h^{(k)}$ it is not the truncation error and cancellation error in $\bar{g}^{(k)}$ that needs to be balanced, as can be seen from the use made of $g^{(k)}$ within the DFP algorithm. It is used, for instance, to form the vector $y^{(k)}$ where

$$y^{(k)} = g^{(k+1)} - g^{(k)}.$$

If $y^{(k)}$ is approximated by $\bar{y}^{(k)}$ where

$$\bar{y}_i^{(k)} = \bar{g}_i^{(k+1)} - \bar{g}_i^{(k)}, \quad i = 1, \ldots, n,$$

then the truncation error in this approximation is $O(h_i^{(k)2} + h_i^{(k+1)2})$. If, however, $h_i^{(k+1)} = h_i^{(k)} = h_i$, where h_i is a fixed scalar, then the truncation error becomes $O(h_i^3)$. The vector $g^{(k)}$ is also used to calculate $p^{(k)}$. The error introduced into this calculation by replacing $g^{(k)}$ by $\bar{g}^{(k)}$ is not minimized by Stewart's choice of $h^{(k)}$.

Gill and Murray (1971) have adapted their implementation of a quasi-Newton algorithm, described in the previous section, to accept difference approximations to the gradient. The differencing interval $h_i^{(k)} = h_i$ is fixed independent of k. The gradient is approximated using (7.7.1) unless this results either in a gradient which is smaller than some prescribed tolerance or in an ascent direction of search. In either of these eventualities the gradient is re-estimated using (7.7.2). Note that since h_i is fixed this re-estimation requires no more function evaluations than those required had (7.7.2) been used initially.

7.8 Error analysis

Because of rounding error in evaluating $F(x)$ and its derivatives all optimization algorithms can at best find the minimum of some function which is a perturbation of $F(x)$. Let $\bar{x}$ be the computed solution, then $\bar{x}$ is a stationary point of

$$\bar{F}(x) = F(x) - \bar{g}^T x,$$

where $\bar{g}$ is the gradient of $F(x)$ at $\bar{x}$.

If the Hessian matrix at $\bar{x}$, denoted say by $\bar{G}$, is positive definite then $\bar{x}$ is a strong local minimum of $\bar{F}(x)$. Otherwise $\bar{x}$ is a strong local minimum of $\hat{F}(x)$, where

$$\hat{F}(x) = F(x) - \bar{g}^T x + \tfrac{1}{2}(|\bar{\lambda}| + \delta)(x - \bar{x})^T (x - \bar{x}),$$

with $\bar{\lambda}$ the smallest eigenvalue of $\bar{G}$
and δ a very small positive scalar.

An estimate to $\bar{g}$ can be obtained in algorithms which either utilize derivatives or approximate derivatives by finite differences. In methods that utilize second derivatives an estimate to $\bar{\lambda}$ can also be obtained.

7.9 Selection of free input parameters

A computer program to minimize a function will require the user to set a number of input parameters. Some of these parameters will be fixed by the problem to be solved and the computer being employed. Typical of such parameters are the number of variables and wordlength of the computer. Other parameters can be varied even when solving the same problem on the same computer and so we term these free parameters. The following list gives some of the free parameters the user may be required to supply.

1. Final convergence criterion.
2. Linear search convergence criterion.
3. Finite-difference intervals.
4. Estimate of $\overset{*}{x}$.
5. Lower bound or estimate of $F(\overset{*}{x})$.
6. Estimate of the Hessian matrix.
7. Estimates of the range of values the variables can take.

The reason for leaving these parameters to be fixed by the user is to enable him to adapt the program to his particular problem. A poor selection of these parameters could result in the program failing, whereas with a good selection it would have succeeded. It is clear that the user should exercise care when choosing these parameters and heed any advice given by the program's author.

In selecting the final convergence criterion the user is faced with conflicting objectives. One aim is to prevent the program from terminating its run prematurely and the second is to prevent needless work being done in refining the computed solution to an unwanted accuracy. One means of overcoming this difficulty is to run the program with the desired criterion and, if the computed solution looks doubtful, to re-run the program with a stricter criterion.

The selection of the linear search criterion may also be complicated by conflicting objectives. With some algorithms a strict linear search criterion is usually less efficient, but more reliable, than a weak criterion. Again the user could first try a weak criterion then re-run the program with a stricter criterion if the computed solution looks doubtful. A more difficult choice is faced when the optimum linear search criterion varies with the size of problem. This arises because a weak criterion, whilst reducing the total number of function evaluations, usually increases the number of iterations. The correct choice can only be made when the computing effort required to evaluate the function compared with that to compute the direction of search is known. Apart from housekeeping operations the computation of the direction of search may involve the evaluation of derivatives which are not required in the linear search. The reduction in number of iterations resulting from using a strict linear search criterion will be unknown but in our experience is unlikely, except for penalty and barrier functions, to be more than 50%.

It may be thought that the best estimate to $\overset{*}{x}$ is that corresponding to the smallest known function value. While this is often true it can sometimes be a very poor choice. It could be that an algorithm has already failed; consequently, the point corresponding to the lowest known function value is the point at which this algorithm has failed. Usually an algorithm fails because the function is ill-behaved in the neighbourhood of the point of failure. An algorithm is more likely to succeed if the initial estimate to $\overset{*}{x}$ does not lie in a region where the function is ill-behaved.

When the user has no means of arriving at a constructive guess of a required estimate then care should be taken to avoid a false estimate which might mislead the program. Thought should be given to why the program requires the estimate and what action is taken on this information. For example, the bounds on the variables and $F(\overset{*}{x})$ may be required to restrict the initial step to be taken in the linear search procedure. The more likely effect of a misleading estimate is to impair the efficiency of a program but it is possible for a poor program to fail on this account. If no constructive bounds are known then wide bounds should be given for the variables and a strict lower bound given for $F(\overset{*}{x})$.

7.10 Interpretation of computer results

Having run a good program with a sensible choice of parameters and the function and its derivatives programmed correctly, the user will be faced with deciding whether or not to accept the estimate of the solution obtained. It is not in general possible to be absolutely sure that the computed solution obtained using any minimization algorithm is a close approximation to the solution. The confidence placed on taking the correct decision will vary with the method used and will obviously be higher for those methods utilizing derivatives.

Usually published algorithms written for general use do not provide print statements and so these will need to be added by the user. The print statements should be designed so that various levels of output can be specified. The minimum level of output should be sufficient to enable a decision to be taken on whether or not more information is required.

If a strict lower bound on $F(x)$ is known and the solution $\bar{x}$ obtained is such that $F(\bar{x})$ is very close to this bound then clearly $\bar{x}$ is an adequate approximation to $\overset{*}{x}$. A strict lower bound is obviously known when $F(x)$ is a sum of squares, that is

$$F(x) = \sum_{i=1}^{n} f_i^2(x) \geqslant 0.$$

Even for these problems there may be some difficulty in deciding whether the approximation is close enough. For instance, in solving a problem using the DFP algorithm in which a strict lower bound of zero was known, the solution obtained gave

$$F(\bar{x}) = 5 \times 10^{-5}.$$

Solving the same problem using the complementary DFP algorithm the solution obtained was (correct to the figures quoted)

$$F(\bar{x}) = 1 \times 10^{-6}.$$

The DFP solution could easily have been accepted had the only check on the solution been to test whether $F(\bar{x})$ was small.

In this instance the problem arose from an approximation problem where increasing the number of variables gave a closer approximation. A good test on the solution can often be made by solving such problems with different numbers of variables. Let $n_1 < n_2 < \ldots < n_j \ldots$ be such numbers and let the corresponding solutions obtained be $F_1(\bar{x}) \ldots F_j(\bar{x}) \ldots$. If it is known that

$$\lim_{j \to \infty} F_j(\overset{*}{x}) \to 0,$$

we can by observing say $F_1(\bar{x})$, $F_2(\bar{x})$ and $F_3(\bar{x})$ form an opinion as to whether

$$\lim_{j \to \infty} F_j(\bar{x}) \to 0.$$

It may be thought that this procedure is inefficient since it requires solving several minimization problems. Usually, however, the solution for $n = n_1$ can be used to construct a good initial estimate for $n = n_2$ and it may in fact be more efficient to obtain the final solution in this manner.

If no special information is available that can confirm $\bar{x}$ to be a close approximation to the solution the following tests should be applied when this is possible.

TEST	RESULT OF TEST
1. Test the final rate of convergence by constructing a table of $F(x^{(k)}) - F(\bar{x})$.	(i) Rate of convergence is superlinear. (ii) Rate of convergence is linear but fast. (iii) Rate of convergence is linear and slow.
2. Test the magnitude of $\|g(\bar{x})\|$.	(i) $\|g(\bar{x})\| < 10 \times 2^{-t}$. (ii) $10 \times 2^{-t} \leqslant \|g(x)\| \leqslant 2^{-t/2}$. (iii) $\|g(\bar{x})\| > 2^{-t/2}$.
3. Test whether or not the Hessian matrix is positive-definite and determine a bound on κ its condition number.	(i) Positive definite, $\kappa < \|g(\bar{x})\|^{-\frac{1}{2}}$. (ii) Positive definite, $\kappa < 0{\cdot}1\|g(\bar{x})\|^{-1}$. (iii) Indefinite or $\kappa \geqslant 0{\cdot}1\|g(\bar{x})\|^{-1}$.

It is assumed the computation is done on a t bit wordlength computer.

If after applying all the tests the results are always (i) then it is almost certain that $\bar{x}$ is a close approximation to the solution. Conversely, if the results are all (iii) it is very unlikely that $\bar{x}$ is close to a strong local minimum. For those methods for which the Hessian matrix or gradient are unknown the tests can be applied to approximations to these quantities if they are available. The tests need to be more stringent in these cases to allow for bias in the approximation error.

The methods most likely to provide estimates which will satisfy these tests are the second derivative methods and those least likely are the direct search methods. The following list gives a rough guide to the relative merits of the various methods with regard to their ability to satisfy the convergence tests. The higher in the list the better the method.

1. Second derivative methods.
2. Quasi-Newton methods.
3. Conjugate gradient methods.
4. Discrete quasi-Newton methods.
5. Conjugate direction methods.
6. Direct search methods.

If the results of the tests on $\bar{x}$ are inconclusive then there are a number of alternative moves that can be made. The first step should be to re-run the program with $\bar{x}$ as the initial estimate to $\overset{*}{x}$ and the convergence criterion reduced. If this fails to produce any significant change to $\bar{x}$ a new initial estimate to $\overset{*}{x}$ should be tried. It may prove worthwhile to vary the other input parameters and rescale the variables. If after three or four attempts convergence is always to a point close to $\bar{x}$ and the tests still do not confirm this to be a local minimum there is little merit in persisting with this algorithm.

The next step is to try a different method, preferably one higher in the list of merit given earlier. This may involve evaluating first derivatives or second derivatives when they were not previously required. When this is not possible a different type of method requiring the same information as the first method should be used. If after making such a change, all the points found are very close to $\bar{x}$ this is in itself an indication that $\bar{x}$ is a close approximation to $\overset{*}{x}$.

8. A Survey of Algorithms for Unconstrained Optimization

R. FLETCHER

Atomic Energy Research Establishment, Harwell

8.1 Introduction

This chapter will consider routines, and in particular FORTRAN IV subroutines or ALGOL 60 procedures, for finding a local minimum of a function $F(x)$ of several variables $x = (x_1, x_2, \ldots, x_n)$. It will be assumed that the function is differentiable, so that the local minimum x^* satisfies the condition that $g(x^*) = 0$, where $g_i = \partial F/\partial x_i$. The concept of the Hessian matrix G will also be referred to, where $G_{ij} = \partial^2 F/(\partial x_i\, \partial x_j)$. It is impossible (or at least grossly inconvenient) to write a single subroutine which will solve all such optimization problems efficiently, and so a number of methods have been developed according to the special circumstances of any one problem. One such special case arises if $F(x)$ is a sum of squares of $m \geqslant n$ non-linear functions $f_i(x)$, $i = 1, 2, \ldots, m$, that is if

$$F(x) = \sum_{i=1}^{m} [f_i(x)]^2 = f^T f.$$

In this case the $m \times n$ Jacobian matrix J of first derivatives of f is of interest ($J_{ij} = \partial f_i/\partial x_j$), and it relates the vector f to the gradient g through the relationship $g = 2J^T f$. Such problems arise in non-linear least squares data fitting when m is usually much greater than n, or when attempting to solve a system of non-linear algebraic equations $f(x) = 0$, when m is equal to n. Other special cases concern whether or not analytic formulae for evaluating first or second derivatives can be determined, and if so how conveniently. Yet again the size and structure of the problem is important and the latter can sometimes be used to advantage.

In this chapter, I have tried to document some of the implementations of what I consider to be the better optimization methods. One overriding consideration has been to choose routines which are freely available to any user. This has led to three possible sources of material—either through algorithms supplements, authors' papers, or through organizations which have specialized in developing optimization routines. In the latter case, whilst certain organizations have indicated that they will be willing to supply subroutines on request, this of course implies no definite commitment on their part, and is subject to any change in the policy on the part of these organizations. In this chapter the initials CACM and CJ will be used to denote routines appearing in the algorithms supplements of the *Communications of the Association for Computing Machinery*, and the *Computer Journal*, respectively. In each case a routine will be indicated by the appropriate code number, e.g. CACM 315. No worthwhile minimization subroutines have yet appeared in the journal BIT which also has an algorithms supplement.

As for organizations which specialize in developing optimization routines, I shall refer to the following:

National Physical Laboratory: (W. Murray, Division of Numerical Analysis and Computing, NPL, Teddington, Middlesex).

Numerical Optimization Centre (The Director, N.O.C., Hatfield College of Technology, Hatfield, Herts.).

Atomic Energy Research Establishment, Harwell (The Subroutine Librarian, Mathematics Branch, AERE, Harwell, Didcot, Berks.).

Subroutines in these libraries will be referred to by the initials of the organization followed by the identifier of the routine, e.g. NPL: NEWTONMIN. Many of the AEREH routines are issued as UKAEA Research Group reports which can be obtained from H.M. Stationery Office at a small charge. Other organizations (National Computer Centre, Imperial Chemical Industries, etc.), also have a number of routines available, but as far as I know these are not free of charge. A second criterion for inclusion has been that the algorithm is reliable. Considerations therefore of whether there is proof (or good reason to think) that convergence will occur and at an ultimately rapid rate, and whether a reasonable amount of satisfactory practical experience has been accumulated with the method, have also influenced my decision.

The intending user of an optimization routine ought to be particularly careful in the initial setting up of an optimization problem because optimization routines are much more sensitive than say routines for solving sets of linear equations. For instance, avoidance of cancellation errors in the evaluation of the function and derivatives is most important, especially in methods which use this information to build up estimates of higher order derivatives by differences. Another feature which is often overlooked is the importance of scaling the variables adequately, the aim being to modify the function so that both the Hessian matrix G and its inverse are close to a unit matrix I, roughly speaking. The effect of bad scaling in quasi-Newton methods is that information about small eigenvalues can be affected by round-off errors, leading for instance to a matrix, which ought to be positive definite given exact arithmetic, having negative eigenvalues in practice. However, new developments by Gill and Murray (1971) in the NPL series of quasi-Newton methods, alleviate these effects of bad scaling to a large extent. Scaling is also important for a different reason in methods which bias the direction of search towards the steepest descent direction in order to guarantee convergence. This is because the steepest descent direction is not independent of scaling, and is more satisfactory when G is close to I. Methods where scaling is important for this reason will be indicated below.

Some other miscellaneous points ought to be kept in mind when posing the problem. It often happens that it is possible to evaluate first derivatives (g or J) but that this will require a certain amount of extra effort on the part of the user. My experience has been that this extra information is usually extremely valuable, enabling methods to be used which give an order of magnitude improvement in the time taken to solve a problem. Furthermore, the effect of differencing errors is reduced considerably. However, it is almost obligatory that any except the most trivial formula for derivatives be checked by differences.

A similar decision may arise on whether or not to evaluate the Hessian matrix G of second derivatives. My experience is that this information whilst useful, will not give rise to anything like so significant an improvement over methods using just first derivatives. It seems that the updating formulae for G or G^{-1} of the quasi-Newton methods, and the approximation $G \approx 2J^TJ$ used in the non-linear least squares methods, are usually quite satisfactory in maintaining adequate approximations to the Hessian. However, if the derivatives are readily available, or if the approximation $G \approx 2J^TJ$ is leading to slow or non-convergence, or if differencing errors are proving troublesome, then I would recommend the evaluation of second derivatives if possible.

Another point of interest concerns non-linear least squares methods. In this case it often happens that the functions f_i are linear in some of the variables, and this can be taken into account when solving the problem. It is possible to pose the problem in terms of the *non-linear* variables only, the linear variables being obtained at each call of the user subroutine by a linear least squares calculation. This device reduces the number of variables in the optimization problem and so usually reduces the number of iterations required for convergence. Some care, however, is required in calculating derivatives in the modified formulation because of the dependence of F not only directly on the non-linear variables, but indirectly through the linear variables which are dependent on the non-linear variables.

The rest of this chapter concerns the availability of subroutines, together with a few remarks indicating how effective each subroutine is in some of the aspects mentioned in this introduction. Section 8.2 concerns routines for non-linear least squares problems, and Section 8.3 routines for general minimization problems. A notable exception is in routines for minimization under gross errors in function values; no conveniently available routines are known to the author.

8.2 Non-linear least squares

In this section all the methods will be variations upon Newton's method in which the approximation $G \approx 2J^T J$ is used. Although there are dangers in using this approximation, especially in that the ultimate rate of convergence will only be linear except in special circumstances (for instance $f = 0$ at the solution), the approximation usually gives good estimates of the Hessian. When such an approximation is used in conjunction with Newton's method it becomes the Gauss–Newton or Generalized Least Squares method. A line search is usually incorporated to prevent divergence of the iteration. Unfortunately the method can fail because it is possible for the sequence $x^{(k)}$ to converge to a point at which the gradient g is not zero, and at which the Jacobian J is not of full rank. Powell (1970a) gives such an example. However, the methods are adequate in most circumstances, especially when $m \gg n$. There are many such programs, for instance CACM 315 (ALGOL) and HOC: MEANDER. An efficient version when derivatives are not available is given by Powell (1965), available in FORTRAN as AEREH: VAO2A. When $m = n$ the Gauss–Newton method becomes Newton's method for solving non-linear equations, and here my experience has been that non-convergence is comparatively frequent and that a routine with guaranteed convergence should be used.

It is possible to ensure convergence at negligible cost by allowing the possibility of biasing the direction of search towards the steepest descent direction, and I would in general recommend methods which do this. Marquardt (1963) gives one such algorithm which requires the user to calculate derivatives, and quotes a subroutine No. 1428 available through the IBM Share library. However, this routine seems difficult to obtain and is ostensibly available to IBM users only. A FORTRAN routine due to FLETCHER† (1971a) also implements Marquardt's method, but with a few modifications designed to improve efficiency. This routine is also available as AEREH: VAO7A. Automatic scaling is available with these routines. When derivatives are not available then efficient routines are available with a similar philosophy. I would recommend the FORTRAN routines AEREH: VAO5A of Powell (1971) when $m > n$, and AEREH: NSO1A of POWELL (1970a) when $m = n$. No automatic scaling is incorporated with these routines so choice of scaling is important.

† A reference in which the author's name is given in capital letters indicates that a listing of the routine appears explicitly in the article referred to.

A special situation arises when solving non-linear equations ($m = n$), and when the Jacobian matrix J is known to be either a general sparse matrix, or a band matrix (such as when solving non-linear two point boundary value ordinary differential equations by finite difference methods). This special structure enables considerable computational savings to be made and hence allows much larger problems to be solved. In these circumstances Marquardt's method would be unfavourable because forming $J^T J$ would either double the band width or would destroy an arbitrary sparsity pattern. Versions of Powell (1970a) for structured problems have been written in FORTRAN by REID (to appear) and each can be used whether or not derivatives are available. When J is a band matrix then use of AEREH: NSO2A is appropriate, whilst AEREH: NSO3A caters for the general sparse matrix. In the latter case the user can specify the sparsity structure, or have it found automatically for him. A further routine AEREH: NSO4A caters for the special case in which J is band symmetric and furthermore is known to be everywhere positive definite. Use of these routines involves the incorporation of other routines in the Harwell subroutine library for the efficient factorization of sparse or band matrices.

8.3 General minimization

Most of the methods to be discussed in this section will be quasi-Newton methods, which work by storing an approximation to G or G^{-1} and updating this approximation at each iteration according to information obtained on that iteration. There are many updating formulae, but most methods use formulae which retain positive definiteness in the approximation. Two formulae in particular have been used: that in the DFP method (see Fletcher and Powell, 1963) which will be called the DFP formula, and that suggested by Broyden (1970), Fletcher (1970) and by others, sometimes known as the complementary DFP formula. However, Dixon (1971) has shown that for general functions, these formulae give identical results when used with exact line searches. My impression is that there is also little evidence in favour of either formula with the line searches as implemented in practice, so the choice of formula is unlikely to be critical.† When derivatives are not available, quasi-Newton methods in which derivatives are estimated by derivatives work well, and Stewart (1967) suggests one way in which this can be done. This algorithm is implemented in ALGOL by Shirley A. Lill in CJ: 46 using the DFP formula and has been modified by RHEAD (1970) in HOC: VMFPDF. However, I believe the attention paid to truncation

† Recent experience with the NPL series of quasi-Newton routines suggests that the complementary DFP formula is preferable, requiring fewer function evaluations per iteration with accurate line searches and failing less frequently with crude line searches.

errors and to maintaining positive definiteness under round-off in the work of Gill and Murray (1971) to be important, and recommend either of the ALGOL routines NPL: DISDFP or NPL: DISCOMDFP. These algorithms use the DFP formula and complementary DFP formula respectively. The quasi-Newton methods seem to be preferable to the Powell (1964) conjugate direction algorithm available in FORTRAN as AEREH: VAO4A, and which has been use a lot in the past.

When derivatives are available there are many similar algorithms and it is difficult to make any firm recommendations. Another complication in these circumstances lies in the way in which the line search is carried out. Some authors prefer to use only function values in the line search and evaluate derivatives only when correcting the approximation to the Hessian; others prefer to use derivatives in both situations. The economics depend upon the ratio

$$\frac{\text{extra number of operations to compute the vector } g}{\text{number of operations to compute } F}.$$

The former approach is better if this ratio is about n; my experience is that a ratio of 2 or 3 is more typical in practice and hence my preference is for the latter approach. There are a number of algorithms which do this, for instance CACM 251 (Revised) (ALGOL), AEREH: VAO1A (FORTRAN), NPL: DESDFP (ALGOL), and HOC: FPMOD1, all of which use the DFP updating formula. There are also two routines which use the complementary DFP formula, NPL: DESCOMDFP (ALGOL) and HOC: BVM1. A procedure based on using function values in the line search together with the complementary DFP formula has been given by K. Fielding in CACM 387.

Recent research on quasi-Newton methods has concentrated on avoiding the line search problem where possible and such a method is given by Fletcher (1970) which chooses between the DFP formula and its complement on each iteration. This algorithm has performed well in a number of independent tests, and is given in Fletcher (1971b). A similar routine due to Biggs (1970) is available through HOC: VMFUM. Another routine which avoids the line search, and also uses a different updating formula, is due to POWELL (1970b) and available as AEREH: VAO6A, for which convergence and superlinear convergence is guaranteed. However, attention to scaling is important when using this method.

When dealing with large problems for which a square symmetric matrix cannot be stored in the computer, then the method of conjugate gradients should be used and an ALGOL procedure is given by FLETCHER and REEVES (1964). An exception to this may be in problems for which the Hessian G is large but sparse. Although no routines are known which take advantage of such structure, there is clearly scope for quasi-Newton routines which are similar to the non-linear least squares routines referred to in Section 8.2.

Finally, the case in which both first and second derivatives are available is considered, and here an ALGOL procedure based on Murray's method described in Chapter 4 can be recommended, available as NPL: NEWTONMIN. This uses a device for biasing the search direction towards the steepest descent direction so attention to scaling may be important when using this procedure.

APPENDIX

Some Aspects of Linear Algebra Relevant to Optimization

Introduction

This appendix defines a number of terms in linear algebra which may be unfamiliar to the reader. It also states without proof a number of results which are relevant to some of the algorithms. A reader wishing to become more versed in the numerical aspects of linear algebra would be well advised to read the books by Fox (1964), Householder (1964) and Wilkinson (1965).

Definition (1)

A symmetric matrix A is said to be a positive-definite matrix if for any vector y, $\|y\| > 0$,

$$y^T A y > 0.$$

Definition (2)

A symmetric matrix A is said to be a positive-semidefinite matrix if it is not a positive-definite matrix and for any vector y, $\|y\| > 0$,

$$y^T A y \geqslant 0.$$

Definition (3)

Let A be an $n \times n$ symmetric matrix, then there exists n orthonormal vectors $v_1, \ldots, v_n$ and n scalars $\lambda_1, \ldots, \lambda_n$ such that

$$Av_i = \lambda_i v_i, \quad i = 1, \ldots, n.$$

The vector v_i is an eigenvector of A and λ_i is its associated eigenvalue.

Definition (4)

A set of vectors $a_1, a_2, \ldots, a_n$ is said to be linearly independent if

$$\sum_{j=1}^{n} \beta_j a_j = 0,$$

implies

$$\beta_j = 0, \quad j = 1, \ldots, n.$$

Definition (5)

The rank of a matrix A is equal to the maximum number of linearly independent rows.

Definition (6)

The space spanned by a set of vectors is the space generated by all linear combinations of those vectors.

Definition (7)

The range of a matrix A, denoted say by $R(A)$, is the space spanned by the columns of A. If $y \in R(A)$ then there exists a vector x such that

$$y = Ax.$$

Definition (8)

The null space of a matrix A, denoted say by $N(A)$, is the space spanned by the vectors orthogonal to the columns of A. If $y \in R(A)$ and $w \in N(A)$ then $y^T w = 0$.

Definition (9)

The pseudo-inverse of an $m \times n$ matrix A is defined as the $n \times m$ matrix X which satisfies the four equations

(1) $AXA = A$.

(2) $XAX = X$.

(3) $(AX)^T = AX$.

(4) $(XA)^T = XA$.

Definition (10)

The condition number of a non-singular matrix A is defined to be κ where

$$\kappa = \|A\| \, \|A^{-1}\|.$$

Definition (11)

The spectral norm of a symmetric non-singular matrix A is defined to be

$$\|A\| = |\lambda_{\max}|,$$

where $\lambda_{\max}$ is the eigenvalue of maximum modulus of A.

Result (1)

Consider the set of equations

$$Ax = b, \tag{A1}$$

where A is an $m \times n$ matrix.

and b is an $m \times 1$ vector.

When a solution exists, a particular solution is given by

$$x = Yb,$$

where Y is any $n \times m$ matrix satisfying the first equation in definition (8). The matrix Y is said to be a generalized inverse of A.

Result (2)

If the equations (A1) have a solution then it follows from result (1) that

$$x = Xb, \tag{A2}$$

where X is the pseudo-inverse of A, is a particular solution. It is the solution whose Euclidean length is a minimum. Moreover, if (A1) does not have a solution then x given by (A2) is the solution of minimum Euclidean length to the problem

$$\underset{x}{\text{minimize}} \; \{(Ax-b)^T(Ax-b)\}.$$

Result (3)

A symmetric matrix of rank r has r non-zero eigenvalues.

Result (4)

A positive-definite matrix has positive eigenvalues.

Result (5)

A positive-semidefinite matrix has non-negative eigenvalues with at least one zero eigenvalue.

Result (6)

If A is an $n \times n$ symmetric matrix whose eigenvalues are $\lambda_1, \ldots, \lambda_n$ with corresponding orthonormal eigenvectors $v_1, \ldots, v_n$ then

$$A = \sum_{i=1}^{n} \lambda_i v_i v_i^T.$$

In addition, if A is non-singular then

$$A^{-1} = \sum_{i=1}^{n} \lambda_i^{-1} v_i v_i^T.$$

Result (7)

It follows from definition (10) and (11) that the condition number of a symmetric non-singular matrix under the spectral norm (spectral condition number) is given by

$$\kappa = |\lambda_{\max}/\lambda_{\min}|,$$

where λ_{max} is the eigenvalue of largest modulus of A
and λ_{min} is the eigenvalue of smallest modulus.

Result (8)

If A is an $n \times n$ non-singular matrix and x and y are two $n \times 1$ column vectors such that $A + xy^T$ is non-singular then

$$(A+xy^T)^{-1} = A^{-1} - \frac{A^{-1}xy^TA^{-1}}{1+y^TA^{-1}x}.$$

This is often referred to as Householder's modification rule.

References

Abbott, J. P. (to appear). Factors affecting the stability of methods of the Davidon–Fletcher–Powell type. *In* F. Lootsma (ed.) "Numerical Methods for Non-linear Optimization". Academic Press, London and New York.

Bandler, J. W. and McDonald, P. A. (1969). "Optimization of microwave networks by razor search". IEEE Transactions on Microwave Theory and Techniques, MIT-17, 552–562.

Bard, Y. (1968). On a numerical instability of Davidon-like methods. *Maths Comput.* **22,** 665–666.

Bard, Y. (1970). Comparison of gradient methods for the solution of non-linear parameter estimation problems. *SIAM Num. Anal.* **7,** 157–186.

Barnes, J. G. P. (1965). An algorithm for solving non-linear equations based on the secant method. *Comput. J.* **8,** pp. 66–72.

Bartels, R. H., Golub, G. H. and Saunders, M. A. (1970). "Numerical techniques in mathematical programming", presented at the MRC Symposium on "Non-linear programming", Madison, Wisconsin.

Bennett, J. M. (1965). Triangular factors of modified matrices. *Num. Math.* **7,** 217–221.

Biggs, M. C. (1970). "A new variable metric technique taking account of non-quadratic behaviour of the objective function". Hatfield Polytechnic Numerical Optimization Centre, Report No. 17.

Box, G. E. P. (1957). Evolutionary Operation: a method for increasing industrial productivity. *Appl. Statist.* **6,** 81–101.

Box, M. J. (1966). A comparison of several current optimization methods, and the use of transformations in constrained problems. *Comput. J.* **9,** 67–77.

Box, M. J., Davies, D. and Swann, W. H. (1969). "Non-linear optimization techniques". ICI Monograph 5. Oliver and Boyd, Edinburgh.

Brent, R. P. (1971). Algorithms for finding zeros and extrema of functions without calculating derivatives. Stanford Univ. Comp. Sci. Dept. Rep. Stan–CS–71–198.

Brown, K. and Conte, S. (1967). "The solution of simultaneous non-linear equations". Proc. 22 Nat. Conf. ACM. Thomson Book Co., Washington DC, 111–114.

Brown, K. M. and Dennis, J. E. (1971). "New computational algorithms for minimizing a sum of squares of non-linear functions". Yale University Report 71–6.

Broyden, C. G. (1965). A class of methods for solving non-linear simultaneous equations. *Maths. Comput.* **19,** 577–593.

Broyden, C. G. (1967). Quasi-Newton methods and their application to function minimization. *Maths. Comput.* **21,** 368–381.

Broyden, C. G. (1970a). The convergence of single rank quasi-Newton methods. *Maths Comput.* **24,** 365–382.

Broyden, C. G. (1970b). The convergence of a class of double-rank minimization algorithms part 1 and part 2, *J. Inst. Maths Applics* **6,** 76–90, 222–231.

Broyden, C. G. and Hart, W. E. (1970). "A new algorithm for constrained optimization". Presented at the 7th Mathematical Programming Symposium, The Hague, The Netherlands.

Broyden, C. G., Dennis, J. E. and Moré, J. J. (to appear). On local convergence properties of quasi-Newton methods.

Businger, P. and Golub, G. H. (1965). Linear least squares pollution by Householder transformations. *Num. Math.* **7,** 269–276.

Carroll, C. W. (1961). The created response surface technique for optimizing non-linear restrained systems. *Ops Res.* **9,** 169–184.

Davidon, W. C. (1959). "Variable metric method for minimization". AEC Research and Development Report, ANL-5990 (Rev).

Davidon, W. C. (1968). Variance algorithms for minimization. *Comput. J.* **10,** 406–410.

Davies, D. and Swann, W. H. (1969). Review of constrained optimization. *In* R. Fletcher (ed.) "Optimization". Academic Press, London and New York. 187–202.

Dennis, J. E. (1970). On the convergence of Newton-like method. *In* P. Rabinowitz (ed.) "Numerical Methods for the Solution of Non-linear Equations". Gordon and Breach.

Dennis, J. E. (1971). On the convergence of Broyden's method for non-linear systems of equations. *Maths Comput.* **25,** 559–567.

Dennis, J. E. (in press). On some methods based on Broyden's secant approximation to the Hessian. *In* F. Lootsma (ed.) "Numerical Methods for Non-linear Optimization". Academic Press, London and New York.

Dickinson, A. W. (1964). Non-linear optimization: some procedures and examples, Proc. 19th ACM National Conference, pp. E1.2/1–E1.2/8.

Dixon, L. C. W. (1971). "Variable matrix algorithms: Necessary and sufficient conditions for identical behaviour on non-quadratic functions". NOC Technical Report 26.

Emery, F. E. and O'Hagan, M. O. (1966). "Optimal design of matching networks for microwave transistor amplifiers". IEEE Transactions on Microwave Theory and Techniques *MIT*-14, 696–698.

Fiacco, A. V. and McCormick, G. P. (1964). Computational algorithm for the sequential unconstrained minimization technique for non-linear programming. *Mgmt Sci.* **10,** 601–617.

Fiacco, A. V. and McCormick, G. P. (1966). Extensions of SUMT for non-linear programming: equality constraints and extrapolation. *Mgmt Sci.* **12,** 816–829.

Fiacco, A. V. and McCormick, G. P. (1968). "Non-linear programming: sequential unconstrained minimization techniques". John Wiley, New York.

Fletcher, R. (1965). Function minimization without evaluating derivatives—a review. *Comput. J.* **8,** 33–41.

Fletcher, R. (1970a). A class of methods for non-linear programming with terminal and convergence properties. *In* J. Abadie (ed.) "Integer and non-linear programming". North-Holland, Amsterdam. 157–175.

Fletcher, R. (1970b). A new approach to variable metric algorithms. *Comput. J.* **13,** 317–322.

Fletcher, R. (1971a). "A modified Marquardt subroutine for non-linear least squares". UKAEA Research Group Report, AERE R. 6799.

Fletcher, R. (1971b). "A survey of algorithms for unconstrained optimization". Report TP 456, AERE, Harwell.

Fletcher, R. and Lill, S. A. (1970). "A class of methods for non-linear programming—Part II, computational experience". Presented at the MRC Symposium on Non-linear programming, Madison, Wisconsin.

Fletcher, R. and McCann, A. P. (1969). Acceleration techniques for non-linear programming. *In* R. Fletcher (ed.) "Optimization". Academic Press, London and New York. 203–214.

Fletcher, R. and Powell, M. J. D. (1963). A rapidly convergent descent method for minimization. *Comput. J.* **6,** 163–168.

Fletcher, R. and Reeves, C. M. (1964). Function minimization by conjugate gradients. *Comput. J.* **7,** 149–154.

Fox, L. (1964). An introduction to numerical linear algebra. Oxford University Press.

Gauss, K. F. (1809). "Theoria motus corporum coelistiam, Werke". **7,** 240–254.

Gill, P. E. and Murray, W. (1971). Quasi-Newton methods for unconstrained optimization. *Natn. phys. Lab. Math.* **97.**

Gill, P. E., Murray, W. and Pitfield, R. A. (1972). The implementation of two quasi-Newton algorithms for unconstrained optimization. *Natn. phys. Lab. DNAC* **11.**

Goldfarb, D. (1970). A family of variable-metric methods derived by variational means. *Maths Comput.* **24,** 23–26.

Greenstadt, J. L. (1967). On the relative inefficiencies of gradient methods. *Maths Comput.* **21,** 360–367.

Greenstadt, J. L. (1970). Variations of variable metric methods. *Maths Comput.* **24,** 1–22.

Gue, R. L. and Thomas, M. E. (1968). "Mathematical methods in operational research". Macmillan, New York.

Haarhoff, P. C. and Buys, J. D. (1970). A new method for the optimization of a non-linear function subject to non-linear constraints. *Comput. J.* **13,** 178–184.

Hartley, H. O. (1961). The modified Gauss–Newton method for the fitting of non-linear regression functions by least squares. *Technometrics* **3,** 269–280.

Hestenes, M. R. and Stiefel, E. (1952). Methods of conjugate gradients for solving linear systems. *J. Res. natn. Bur. Stand.* **49,** 409–436.

Hestenes, M. R. (1969). Multiplier and gradient methods. *In* L. A. Zadeh, L. W. Neustadt and A. V. Balakrishnan (eds.) "Computing methods in optimization problems". **2.** Academic Press, London and New York. 143–163.

Himmelblau, D. M. (in press). A uniform evaluation of unconstrained optimization techniques. *In* F. Lootsma (ed.) "Numerical Methods for Non-linear Optimization". Academic Press, London and New York.

Hooke, R. and Jeeves, T. A. (1961). "Direct Search" solution of numerical and statistical problems. *J. Ass. comput. Mach.* **8,** 212–229.

Householder, A. S. (1964). The theory of matrices in numerical analysis. Blaisdell Publishing Co., New York.

Huang, H. U. (1970). Unified approach to quadratically convergent algorithms for function minimization. *JOTA* **5,** 405–423.

Huang, H. U. and Levy, A. V. (1970). Numerical experiments on quadratically convergent algorithms for function minimization. *JOTA* **6,** 269–282.

Jarratt, P. (1970). A review of methods for solving non-linear algebraic equations in one variable. *In* P. Rabinowitz (ed.) "Numerical methods for non-linear algebraic equations", Gordon and Breach. 1–26.

Jones, A. (1970). Spiral—a new algorithm for non-linear parameter estimating using least squares. *Comput. J.* **13,** 301–308.

Kiefer, J. (1957). Optimal sequential search and approximation methods under minimum regularity conditions. *SIAM J. appl. Maths* **5,** 105–136.

Kowalik, J. and Osborne, M. R. (1968). "Methods for unconstrained optimization problems". Elsevier, New York.

Krolak, P. and Cooper, L. (1963). An extension of Fibonaccian search to several variables. *Commun. Ass. comput. Mach.* **6,** 639–641.

Kwakernaak, H. and Strijbos, R. C. W. (1970). "Extremization of functions with equality constraints". Paper presented at the VII International Symposium of Mathematical Programming, The Hague.

Levenberg, K. (1944). A method for the solution of certain non-linear problems in least squares. *Q. appl. Math.* **2,** 164–168.

Lootsma, F. A. (1970). "Boundary properties of penalty functions for constrained minimization". Thesis, University of Eindhoven, The Netherlands.

Maddison, R. N. (1966). A procedure for non-linear least squares refinement in adverse practical conditions. *J. Ass. comput. Mach.* **13,** 124–134.

Marquardt, D. W. (1963). An algorithm for least squares estimation of non-linear parameters. *SIAM J.* **11,** 431–441.

Matthews, A. and Davies, D. (1969). "A comparison of modified Newton methods for unconstrained optimization". ICI Ltd., Central Management Services research note.

Matthews, A. and Davies, D. (1971). A comparison of modified Newton methods for unconstrained optimization. *Comput. J.* **14,** 293–294.

McCormick, G. P. and Pearson, J. D. (1969). Variable metric methods and unconstrained optimization. *In* R. Fletcher (ed.) "Optimization". Academic Press, London and New York.

Miele, A., Cragg, E. E., Iyer, I. I. and Levy, A. V. (1971). Use of the augmented penalty function in Mathematical Programming, **8,** 115–130.

Murray, W. (1969). An algorithm for constrained minimization. *In* R. Fletcher (ed.) "Optimization". Academic Press, London and New York. 247–258.

Murray, W. (1972). The relationship between the approximate Hessian matrices generated by a class of quasi-Newton methods. *Natn. phys. Lab. DNAC* **12.**

Murtagh, B. A. and Sargent, R. W. H. (1970). A constrained minimization method with quadratic convergence. *In* R. Fletcher (ed.) "Optimization". Academic Press, London and New York.

Nelder, J. A. and Mead, R. (1965). A simplex method for function minimization. *Comput. J.* **7,** 308–313.

Ortega, J. M. and Rheinboldt, W. C. (1970). "Iterative solution of non-linear equations in several variables". Academic Press, New York and London.

Palmer, J. R. (1969). An improved procedure for orthogonalizing the search vectors in Rosenbrock's and Swann's direct search optimization methods. *Comput. J.* **12,** 69–71.

Pearson, J. D. (1969). Variable metric methods of minimization. *Comput. J.* **12,** 171–178.

Peckham, G. (1970). A new method for minimizing a sum of squares without calculating gradients. *Comput. J.* **13,** 418–420.

Polak, E. and Ribiere, G. (1969). "Note sur la convergence de methodes des directions conjugées". University of California, Berkeley, Dept of Electrical Engineering and Computer Sciences, working paper.

Powell, M. J. D. (1962). An iterative method for finding stationary values of a function of several variables. *Comput. J.* **5,** 147–151.

Powell, M. J. D. (1964). An efficient method of finding the minimum of a function of several variables without calculating derivatives. *Comput. J.* **7,** 155–162.

Powell, M. J. D. (1965). A method for minimizing a sum of squares of non-linear functions without calculating derivatives. *Comput. J.* **7,** 303–307.

Powell, M. J. D. (1968). On the calculation of orthogonal vectors. *Comput. J.* **11,** 302–304.

Powell, M. J. D. (1969a). A method for non-linear constraints in minimization problems. *In* R. Fletcher (ed.) "Optimization". Academic Press, London and New York. 283–293.

Powell, M. J. D. (1969b). A theorem on rank one modification to a matrix and its inverse. *Comput. J.* **12,** 288–290.

Powell, M. J. D. (1970a). A hybrid method for non-linear equations. *In* P. Rabinowitz (ed.) "Numerical methods for non-linear algebraic equations". Gordon and Breach. 87–114.

Powell, M. J. D. (1970b). A FORTRAN subroutine for solving systems of non-linear algebraic equations. *In* P. Rabinowitz (ed.) "Numerical methods for non-linear algebraic equations". Gordon and Breach. 115–161.

Powell, M. J. D. (1970c). A FORTRAN subroutine for unconstrained minimization requiring first derivatives of the objective function. UKAEA Research Group Report, AERE R. 6469.

Powell, M. J. D. (1970d). A new algorithm for unconstrained optimization. Report TP 393, AERE, Harwell.

Powell, M. J. D. (1971). AERE Harwell library subroutine VAO5A, report in preparation.

Reid, J. K. (to appear), Fortran subroutines for the solution of sparse systems of non-linear equations.

Rhead, D. G. (1970). A modification to Lill's implementation of Stewart's method, Hatfield Polytechnic, Numerical Optimization Centre, Report 18.

Rhead, D. G. (1971). "Some numerical experiments on Zangwill's method for unconstrained minimization". University of London, Inst. of Computer Sci., Working Paper ICSI 319.

Rosen, J. B. (1960). The gradient projection method for non-linear programming, part 1—"linear constraints". *SIAM J.* **8,** 181–217.

Rosenbrock, H. H. (1960). An automatic method for finding the greatest or least value of a function. *Comput. J.* **3,** 175–184.

Rutishauser, H. (1959). *In* M. Engeli *et al.* (eds.) "Theory of gradient methods, in refined iterative methods for computation of the solution and the eigenvalues of self-adjoint boundary value problems". Birkhäuser, Basel.

Samanski, V. (1967). On the modification of the Newton method. (Russian) Ukrain. *Math. Z.* **19,** 218–227.

Shah, B. V., Buehler, R. J., Kempthorne, O. (1964). Some algorithms for minimizing a function of several variables. *J. SIAM.* **12,** 74–92.

Shanno, D. F. (1970). Conditioning of quasi-Newton methods for function minimization. *Maths Comput.* **24,** 647–656.

Smith, C. S. (1962). "The automatic computation of maximum likelihood estimates". NCB Sc. Dept., Report SC 846/MR/40.

Spang, H. A. (1962). A review of minimization techniques for non-linear functions. *SIAM Rev.* **4,** 343–365.

Spendley, W., Hext, G. R. and Himsworth, F. R. (1962). Sequential application of simplex designs in optimization and Evolutionary Operation. *Technometrics* **4,** 441–461.

Spendley, W. (1969). Non-linear least squares fitting using a modified simplex minimization method. *In* R. Fletcher (ed.) Academic Press, London and New York. 259–270.

Swann, W. H. (1964). "Report on the development of a new direct search method of optimization". ICI Ltd. Central Instrument Research Laboratory Research Note 64/3.

Stewart, G. W. (1967). A modification of Davidon's minimization method to accept difference approximation of derivatives. *J. Ass. comput. Math.* **14,** 72–83.

Wilkinson, J. H. (1963). "Rounding errors in algebraic processes". Her Majesty's Stationery Office, London.

Wilkinson, J. H. (1965). "The algebraic eigenvalue problem". Oxford University Press, London.

Wolfe, P. (1959). The secant method for simultaneous non-linear equations. *Ass. comput. Math. Commun.* **2,** (12), pp. 12–13.

Wolfe, P. (1967). "Another variable metric method". Working paper.

Wolfe, P. (1969). Convergence conditions for ascent methods. *SIAM Rev.* **11,** 226–235.

Zangwill, W. I. (1967). Minimizing a function without calculating derivatives. *Comput. J.* **10,** 293–296.

Zoutendijk, G. (1970). *In* J. Abadie (ed.) "Non-linear programming—computational methods, in Integer and non-linear programming". North-Holland, Amsterdam.

Author Index

Numbers in italics refer to the Reference pages where the references are given in full.

A

Abbott, J. P., 103, *135*

B

Bandler, J. W., 20, *135*
Bard, Y., 29, 33, 35, 61, 93, 103, 114, *135*
Barnes, J. G. P., 39, 40, 97, *135*
Bartels, R. H., 40, 42, *135*
Bennett, J. M., 40, *135*
Biggs, M. C., 128, *135*
Box, G. E. P., 23, *135*
Box, M. J., 15, 42, *135*
Brent, R. P., 112, 135
Brown, K., 88, *135*
Brown, K. M., 104, *135*
Broyden, C. G., 39, 40, 41, 50, 90, 96, 97, 102, 103, 105, 127, *135*, *136*
Buehler, R. J., 83, *139*
Businger, P., 69, *136*
Buys, J. D., 53, *137*

C

Carroll, C. W., 43, 44, *136*
Conte, S., 88, *135*
Cooper, L., 15, *138*
Cragg, E. E., 53, *138*

D

Davidon, W. C., 49, 90, 99, 103, *136*
Davies, D., 15, 42, 63, *135*, *136*, *138*
Dennis, J. E., 94, 97, 101, 102, 104, *135*, *136*
Dickinson, A. W., 42, *136*
Dixon, L. C. W., 102, 127, *136*

E

Emery, F. E., 20, *136*

F

Fiacco, A. V., 44, 45, 46, 47, 48, 63, *136*
Fletcher, R., 31, 49, 53, 54, 55, 77, 78, 81, 90, 94, 95, 103, 126, 127, 128, *136*, *137*
Fox, L., 131, *137*

G

Gauss, K. F., 97, *137*
Gill, P. E., 12, 91, 113, 114, 117, 125, 128, *137*
Goldfarb, D., 90, 103, *137*
Golub, G. H., 40, 42, 69, *135*, *136*
Greenstadt, J. L., 59, 90, *137*
Gue, R. L., 5, *137*

H

Haarhoff, P. C., 53, *137*
Hart, W. E., 50, 105, *136*
Hartley, H. O., 34, *137*
Hestenes, M. R., 52, 79, 85, *137*
Hext, G. R., 24, *139*
Himmelblau, D. M., 96, *137*
Himsworth, F. R., 24, *139*
Hooke, R., 18, *137*
Householder, A. S., 131, *137*
Huang, H. U., 98, 102, 104, *137*

I

Iyer, I. I., 53, *138*

J

Jarratt, P., 37, *137*
Jeeves, T. A., 18, *137*
Jones, A., 36, 41, *137*

K

Kempthorne, O., 83, *139*
Kiefer, J., 9, *138*
Kowalik, J., 33, 46, *138*
Krolak, P., 15, *138*
Kwakernaak, H., 105, *138*

L

Levenberg, K., 34, 138
Levy, A. V., 53, 102, 104, *137*, *138*
Lill, S. A., 53, 55, *137*
Lootsma, F. A., 45, *138*

M

McCann, A. P., 49, *137*
McCormick, G. P., 44, 45, 46, 47, 48, 63, 82, *136*, *138*
McDonald, P. A., 20, *135*
Maddison, R. N., 33, *138*
Marquardt, D. W., 34, 35, 126, *138*
Matthews, A., 63, *138*
Mead, R., 26, *138*
Miele, A., 53, *138*
Moré, J. J., 102, *136*
Murray, W., 12, 49, 91, 113, 114, 117, 125, 128, *137*, *138*
Murtagh, B. A., 90, 99, 116, *138*

N

Nelder, J. A., 26, *138*

O

O'Hagan, M. O., 20, *136*
Ortega, J. M., 36, *138*
Osborne, M. R., 33, 46, *138*

P

Palmer, J. R., 23, *138*
Pearson, J. D., 82, 85, 90, 93, 94, 98,
Peckham, G., 38, 39, *138*
Pitfield, R. A., 12, 113, *137*
Polak, E., 82, *138*
Powell, M. J. D., 23, 31, 34, 35, 39, 40, 41, 51, 52, 77, 83, 90, 94, 100, 101, 102, 103, 112, 126, 127, 128, *137*, *138*, *139*

R

Reeves, C. M., 81, 128, *136*, *137*
Reid, J. K., 127, 139
Rhead, D. G., 78, 112, 127, *139*
Rheinboldt, W. C., 36, *138*
Ribiere, G., 82, *138*
Rosen, J. B., 39, 40, *139*
Rosenbrock, H. H., 21, *139*
Rutishauser, H., 84, *139*

S

Samanski, V., 88, *139*
Sargent, R. W. H., 90, 99, 116, *138*
Saunders, M. A., 40, 42, *135*
Shah, B. V., 83, *139*
Shanno, D. F., 103, *139*
Smith, C. S., 76, *139*
Spang, H. A., 14, *139*
Spendley, W., 24, 38, *139*
Stewart, G. W., 116, 127, *140*
Stiefel, E., 79, *137*
Strijbos, R. C. W., 105, *138*
Swann, W. H., 15, 22, 42, *135*, *136*, *140*

T

Thomas, M. E., 5, *137*

W

Wilkinson, J. H., 60, 63, 113, 131, *140*
Wolfe, P., 34, 37, 97, 99, *140*

Z

Zangwill, W. I., 78, *140*
Zoutendijk, G., 84, *140*

Subject Index

A

ALGOL 60, 123

B

Bisection, 8
Bounded deterioration, 94

C

Cancellation error, 109, 116
Cholesky factorization, 60, 62, 64
Condition number, 121, 132
Conjugate directions, 6, 77, 113
Conjugate gradients, 75, 78
Conjugate vectors, 74
Constraints, 30, 42, 43, 104
Convergence criterion, 118

E

Eigenvalue, 59, 61, 123, 131, 132
Eigenvector, 59, 61, 131, 132
Extrapolation, 49

F

Fibonacci search, 8, 15
Finite-difference approximation, 116
FORTRAN IV, 123
Function
  approximation, 10
  convex, 6, 7, 58
  of a single variable, 7
  unimodal, 15
Functionals, 12

G

Gaussian elimination, 63
Generalized inverse, 133
Golden section, 8

H

Hessian matrix, 4, 57, 70, 74, 87, 123

I

Interpolation, 110

J

Jacobean matrix, 69, 87, 123

L

Lagrange
  multipliers, 105
  parameters, 49, 50
Least squares, 31, 36, 69, 123
Linear algebra, 131
Linear dependent, 78
Linear independent, 131
Linear search, 8, 17, 23, 33, 39, 80, 110, 119
Line search, 74, 77, 83, 88, 97, 128

M

Matrix
  null space, 132
  positive-definite, 5, 59, 74, 113, 131
  positive semidefinite, 4, 32, 54, 87, 131
  range, 132
  rank, 132
Minimum
  absolute, 2
  global, 2
  improper relative, 2
  proper relative, 2
  strong local, 1, 2, 3
  strong relative, 2
  weak local, 1, 2, 3

N

Nonlinear equations, 30, 87, 97, 100, 123, 127

O

Orthogonal directions, 21, 22, 77

P

Penalty function, 30, 43, 103
Principle axis, 17
Projection matrix, 101
Pseudo inverse, 54, 132

Q

Quadratic function, 5, 58, 73, 93
Quadratic termination, 77, 94, 102, 115

R

Rounding error, 82, 88, 108, 109, 115

S

Saddle point, 4, 68
Scaling, 124
Simplex, 24, 26
Spectral norm, 132
Stationary point, 4
Steepest descent, 34, 42, 57, 68, 83, 125, 129
Sum of squares, 29, 31, 35, 100, 123

T

Transformation of variables, 43, 110
Truncation error, 88, 116, 128